SALTERS ADVANCED CHEMISTRY

Revise

AS

Chemistry for Salters

Lesley Johnston, Dave Newton,
Chris Otter, Kay Stephenson,
Alasdair Thorpe

www.heinemann.co.uk

✓ Free online support
✓ Useful weblinks
✓ 24 hour online ordering

01865 888080

OCR RECOGNISING ACHIEVEMENT   Heinemann

Official Publisher Partnership

Heinemann is an imprint of Pearson Education Limited, a company incorporated in
England and Wales, having its registered office at Edinburgh Gate, Harlow, Essex, CM20 2JE.
Registered company number: 872828

www.heinemann.co.uk

Heinemann is a registered trademark of Pearson Education Limited

Text © Ann Daniels, Lesley Johnston, Dave Newton, Alasdair Thorpe, University of York 2008

First published 2004
This edition published 2008

12
10 9 8 7

British Library Cataloguing in Publication Data is available from the British Library on request.

ISBN 978 0 435631 54 3

Edited by Tony Clappison
Designed, produced, illustrated and typeset by Wearset Limited, Boldon, Tyne and Wear
Original illustrations © Pearson Publishers Oxford Limited 2008
Cover design by Wearset Limited, Boldon, Tyne and Wear
Cover photo/illustration © NASA/Science Photo Library
Printed in Malaysia, CTP-KHL

Websites
The websites used in this book were correct and up-to-date at the time of publication. It is essential
for tutors to preview each website before using it in class so as to ensure that the URL is still
accurate, relevant and appropriate. We suggest that tutors bookmark useful websites and consider
enabling students to access them through the school/college intranet.

Contents

Unit F331 Chemistry for Life

Unit F332 Chemistry of Natural Resources

Exam hints and tips

Here are some general points that apply to both the external exams for Unit F331 (Chemistry for Life) and Unit F332 (Chemistry of Natural Resources) written examination papers.

1. **All the questions are structured.** This means that you are given a 'stem' of information that provides the context (the *Storyline*) for the question. This is followed by a series of part-questions. It can be quite helpful to underline key pieces of information as you read through the stem.

2. **The questions are designed so that you work through them in stages.** Answers to the part-questions are often linked together and are linked to the information you are given. Work through the questions in order – don't cherry pick.

3. **Part-questions are linked by the context** – not by chemical topic. You will need to dip into several different parts of your knowledge in one question.

4. **Contexts will be a mixture of familiar ones from the *Storylines* and unfamiliar ones.** Don't panic if the context is unfamiliar – the chemistry you are being asked about *will* be familiar.

5. **The Data Sheet provides additional information not given in the question.** Make sure you are familiar with what is on the Data Sheet – and remember to use it!

6. **All questions are compulsory – you have no choice.** Try to answer every question. It is better to make a sensible guess, which could score you 1 mark, than to put nothing at all. You cannot get negative marks. You have nothing to lose by guessing – so use your chemical common sense!

7. **Examiners use an agreed mark scheme.** The marks are given for very specific points, so you must be precise and use language accurately. Credit will *always* be given for correct chemistry, clearly explained or described, that answers the question that has been asked.

8. **Some questions will require a knowledge of applications of chemistry and the work of chemists.** This means that it is a good idea to have read and made notes on the *Storylines*, particularly those sections listed in the 'Check your knowledge and understanding' activities for each teaching module.

9. **The first question in an exam is designed to be quite straightforward** to help settle you into the exam. So, you are advised to do the questions in the order they are set.

Here are some tips for answering question papers:

- Answer what is asked.
- Give a sign with oxidation states: a '+' or '−' before the number (e.g. +3, −5).
- Give state symbols in chemical equations only when requested. They are often asked for in reactions where there is a change of state.
- Use chemical language correctly (e.g. make sure you know the difference between atom, molecule and ion, between chlorine and chloride, and between hydroxide and hydroxyl).

Types of examination questions

'Explain' questions

These ask you to describe in your own words, in a logical order, the chemical reasons for a fact, observation or suggestion. These part-questions can often require some extended writing. Some will be used to assess the quality of your written communication – you will always be told if this is the case.

Economic, environmental and social questions

These questions can sometimes sound vague, but they require specific answers. You must be careful to give an answer in terms of what you have learnt in your AS chemistry course. Use the information you have been given in the question.

Writing chemical formulae questions

You will be asked to draw chemical formulae for organic molecules.

- A **full structural formula** means that you must show every bond in the molecule. A common error is to miss some of the hydrogen atoms. Alcohol groups should be written –O–H, not –OH. You do not need to draw out all the bonds in benzene though – use the usual hexagon with a circle inside.

- If you are asked simply to draw the **structure** of a molecule, be guided by the style of similar molecules in the question. You do not need to draw out every bond, but the functional groups should be clear and the structure unambiguous.

- A **skeletal formula** shows the carbon skeleton as angled lines in a zig-zag – carbon atoms are assumed to be at the ends of each line. Functional groups then stand out very clearly. Hydrogen atoms are shown only on functional groups (not those directly on the carbon skeleton).

- Dot–cross formulae should show lone (non-bonding) pairs of electrons as well as bonding electrons.

Calculation questions

Always show your working. Examiners use 'consequential marking' where possible to give credit for correct working, even if your final answer is wrong, so make sure you include everything. Even simple things, like calculating numbers of moles or relative molecular masses, can get you a mark.

Here are some other calculation tips:

- Write your final answer on the answer line.

- Always give units (unless already present) and a sign if appropriate (a sign is *always* needed for enthalpy changes).

- Always quote your final result to the same number of significant figures as the data with the lowest number of significant figures quoted in the question (usually 2 or 3 significant figures).

- Don't give the answer only, show your working out.

- Always think about whether your answer is sensible (e.g. $\Delta H_c^{\ominus}$ values are negative).

Advance notice question

In Unit F332, one question will be based on a relevant chemical article. This will be available several weeks before the examination for you to read and to make sure that you understand the chemistry. The examination paper will contain a fresh copy of the article, but there will not be any reading time allocated for this – it is there simply to refer to.

> To get your written communication mark, you must make sure you:
> - present a logical account
> - use correct scientific terms
> - spell these terms correctly.

> Try drawing a full structural formula for pentanoic acid.

> Try drawing the skeletal formula of propanone.

> Try drawing a dot–cross formula for N_2H_4.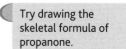

Experimental techniques

There are a number of experimental procedures you are expected to know about in detail for the AS course. The important ones are described below.

Making thermochemical measurements in order to calculate enthalpy changes (see Activity DF2.1)

See pages 20 and 21 for details of the apparatus and calculations used. You also need to be aware that during this type of calculation the following assumptions are usually made:

- the specific heat capacity of solutions being used is the same as that of water ($4.18\,J\,g^{-1}\,K^{-1}$)
- the specific heat capacity of the calorimeter and solids used are disregarded – they are small compared to the overall values being used.

There are other points to consider. Values obtained are often considerably lower than the actual data values quoted – this is due to heat losses to the surroundings or incomplete combustion of a fuel under test. In order to increase the accuracy, heat losses can be minimised by using heat shields to minimise draughts. Also, for example, the spirit burner can be kept as close as possible to the calorimeter – however, care must be taken to ensure that enough oxygen can get to the burner to ensure that complete combustion occurs. Whenever the spirit burner is not in use, the lid must be replaced to prevent evaporation of the fuel between weighings.

Carrying out an acid/alkali titration (see Activity ES5.1)

This technique is used to determine the concentration of a solution by reacting it with another solution of known concentration. When the reaction is complete, it is signalled by the colour change of an indicator. The main stages are:

1 Pipette a known volume of the solution of unknown concentration into a clean, dry conical flask.
2 Place the flask on a white tile – this improves the visibility of any colour changes.
3 Add a few drops of a suitable indicator – do not add too much because the indicator reacts with acids and alkalis and so will interfere with your results.
4 Rinse a clean burette with the solution of known concentration.
5 Fill the burette with this solution.
6 Take an initial reading of the volume from the burette, using the lowest point of the meniscus of the liquid.
7 Swirl the conical flask and add the liquid from the burette $1\,cm^3$ at a time until the expected colour change occurs. Keep swirling to mix the contents of the flask throughout the process.
8 Rinse the conical flask with distilled water and repeat steps 1, 2, 3, 5 and 6. This time run in the solution from the burette up to $1\,cm^3$ before the colour change obtained in step 7. Then, swirling thoroughly, add more solution dropwise until the desired colour change is observed. This gives an accurate result.
9 Repeat step 8 until you get three titre values within $0.1\,cm^3$ of each other. These are called *concordant results*. The average of these titres is used in the calculations. For details of these calculations see pages 36 and 37.

Phenolphthalein and methyl orange are often used as indicators in acid/alkali titrations. Phenolphthalein is magenta in alkali and colourless in acid. Methyl orange is red in acid and yellow in alkali. The end-point for methyl orange is an orange colour.

The volume of liquid delivered from a burette during a titration is known as the *titre*.

Heating under reflux (see Activity PR4.3)

This technique is used to heat volatile reactants together in order to allow a chemical reaction to take place without any of the mixture escaping. This increases yield and reduces the risk of fires. See page vii for advice on how to draw a diagram of this apparatus.

Purifying a liquid using a separating funnel (see Activity ES6.3)

This technique is used to separate two immiscible liquids – often to separate an organic liquid from an aqueous solution. You need to know the relative densities of each layer, so you know which one to keep.

Simple distillation   (see Activity ES6.3)

This technique is used to separate two or more miscible liquids with different boiling points. For details of the apparatus used see the diagram below. When the mixture is heated, the liquid with the lowest boiling point evaporates first.

The mixture is heated until the thermometer shows the vapour temperature to be at the boiling point of the desired liquid. A clean receiver beaker is placed below the condenser to collect the condensed liquid. When the temperature on the thermometer rises above the boiling point of the desired liquid, the distillation process is stopped.

Details of all of these techniques can be found in **Appendix 1** of *Chemical Ideas (third edition)*.

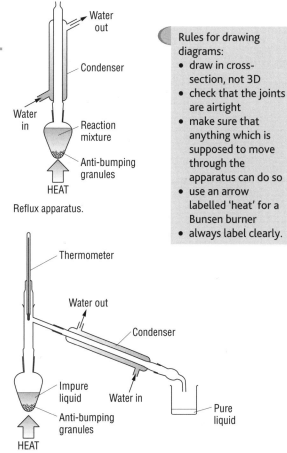

Separating funnel.

Diagrams of experimental equipment

The two diagrams which are most likely to come up are the ones for *heating under reflux* and *simple distillation*.

Heating under reflux

- Note the direction of the water flow.
- Never stopper the top of the reflux condenser.
- Label the reaction mixture.
- Always include anti-bumping granules.

Simple distillation

- Note the direction of the water flow.
- The thermometer bulb must be at *exactly* the spot shown – not higher or lower.
- Always include anti-bumping granules.
- Always include a receiver beaker to catch the condensate.

Describing experimental techniques

Sometimes a question asks you to describe an experimental technique you have met. For example:

The amount of acid in a solution can be measured by titrating a solution with sodium hydroxide solution of known concentration. Describe the main steps involved in such a titration.

You have to decide which are the most important points to include, bearing in mind the number of marks available. A suitable answer might be:
- Pipette out a known volume of the acid solution.
- Add a suitable indicator (e.g. phenolphthalein).
- Titrate, shaking the conical flask, with sodium hydroxide solution from a burette …
  … until the magenta colour becomes permanent.
- Repeat the titration to obtain concordant results.

Reflux apparatus.

Simple distillation apparatus.

> **Rules for drawing diagrams:**
> - draw in cross-section, not 3D
> - check that the joints are airtight
> - make sure that anything which is supposed to move through the apparatus can do so
> - use an arrow labelled 'heat' for a Bunsen burner
> - always label clearly.

> Note that you need a point for each mark awarded in a question.

How to use this revision guide

This revision guide is designed to support the OCR Chemistry B (Salters) AS course, and is valid for the revised specification for first teaching in September 2008, with the first F331 examination taken from January 2009. From June 2009 you may take both F331 and F332 in January or June, or just in June at the end of the AS course.

This guide covers the two written examinations for the AS course – **Unit F331: Chemistry for Life** and **Unit F332: Chemistry of Natural Resources**.

The table shows the scheme of assessment for the AS course. This enables you to see how the examination units and teaching content link together.

Examination Unit	Unit title and teaching modules covered	Duration	Number of marks	Mode of assessment	% weighting of AS
F331	**Chemistry for Life** (Elements of Life and Developing Fuels)	1¼ hours	60	Written examination	30%
F332	**Chemistry of Natural Resources** (Elements from the Sea, The Atmosphere and Polymer Revolution)	1¾ hours	100	Written examination One question will be on an *Advance Notice* article	50%
F333	**Chemistry in Practice**	60 marks: not covered in this revision guide – your school or college will organise this assessment	Practical activities assessed by your teacher	20%

A spider diagram at the start of each module shows how the concepts in the module are related.

Under each topic title in this book there is a summary of the content covered. It allows you to see which sections from *Chemical Ideas, Chemical Storylines* (and in a few cases activities) need to be revised for that specific topic.

Within each teaching topic, the material is divided into short sections of one to three pages. Each section usually revises one section of material from *Chemical Ideas*, and ends with **Quick Check questions** to help you to test your understanding. Shaded marginal boxes provide useful hints, including dealing with common misconceptions and errors.

Also included at the front of this book are sections on **Experimental techniques** and **Exam hints and tips**.

At the end of each Unit are **Practice exam-style questions**, along with helpful marginal comments to assist you in answering them.

All questions (Quick Check questions and Examination-style questions) are provided with full answers.

The Periodic Table of the Elements is provided on the inside front cover.

An index is provided at the end of the book.

Elements of Life (EL)

This module tells the story of the elements of life – what they are, how they originated and how they can be detected and measured. As you work through this module you will cover the following concepts. 'CI' refers to sections in your *Chemical Ideas* textbook. 'Storylines' refers to your *Chemical Storylines* textbook.

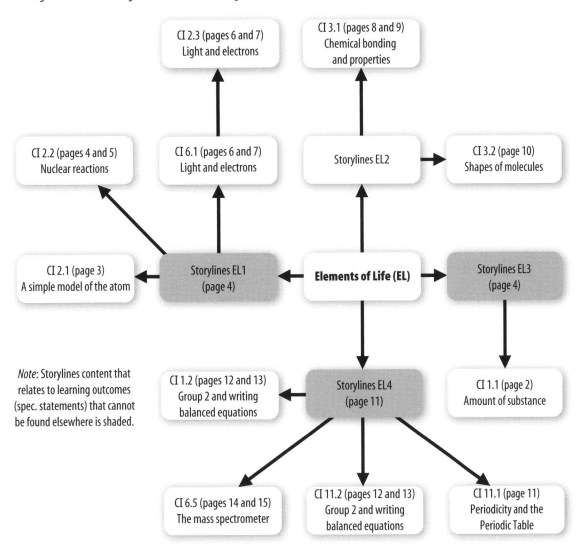

CI 2.3 (pages 6 and 7)
Light and electrons

CI 3.1 (pages 8 and 9)
Chemical bonding
and properties

CI 2.2 (pages 4 and 5)
Nuclear reactions

CI 6.1 (pages 6 and 7)
Light and electrons

Storylines EL2

CI 3.2 (page 10)
Shapes of molecules

CI 2.1 (page 3)
A simple model of the atom

Storylines EL1
(page 4)

Elements of Life (EL)

Storylines EL3
(page 4)

Note: Storylines content that relates to learning outcomes (spec. statements) that cannot be found elsewhere is shaded.

CI 1.2 (pages 12 and 13)
Group 2 and writing
balanced equations

Storylines EL4
(page 11)

CI 1.1 (page 2)
Amount of substance

CI 6.5 (pages 14 and 15)
The mass spectrometer

CI 11.2 (pages 12 and 13)
Group 2 and writing
balanced equations

CI 11.1 (page 11)
Periodicity and the
Periodic Table

As well as the materials covered in this module, you also need to know that our knowledge of the structure of the atom developed in terms of a series of models, which gradually became more sophisticated. Given information, you should be able to interpret these and other examples of such developing models. For more details see **Activity EL1.1**.

Amount of substance

Chemical Ideas 1.1

What you need to know

- 6.02×10^{23} (**Avogadro constant**, given the symbol N_A) is the number of particles in 1 mole of a substance.

- **Relative atomic mass** (A_r) tells you the number of times an atom of an element is heavier than one-twelfth of an atom of ^{12}C. A_r values have no units.

> M_r CO_2: $12.0 + 16.0 + 16.0 = 44.0$

- **Relative formula mass** (M_r) is the sum of the relative atomic masses (A_r) for each atom in the formula. For a simple molecular compound, it is sometimes called **relative molecular mass**. M_r values have no units.

> Ethane has the molecular formula C_2H_6 but its empirical formula is CH_3.

- To obtain **one mole** of a substance, weigh out the A_r or M_r exactly in grams.

- The **empirical formula** tells you the *simplest ratio* of atoms of each element in a compound.

> In an exam, you will be expected to look up the A_r values (*to 1 decimal place*) using the Periodic Table provided in the data sheets.

- The **molecular formula** tells you the *actual number* of atoms of each element in a molecule. This may be either the same as or a whole number multiple of the empirical formula.

To work out the amount in moles of a substance, use amount $= \dfrac{\text{mass (g)}}{A_r}$ for an atomic element; amount $= \dfrac{\text{mass (g)}}{M_r}$ for a molecule or compound.

> If the question gives the percentage composition by mass, then assume the numbers given to be grams and carry on as above. For example, for 16% C, write 16.0 g C.

Relative atomic masses.

Element	A_r
C	12.0
O	16.0
Mg	24.3
H	1.0
N	14.0
S	32.1
Na	23.0
Pb	207.2
Be	9.0

WORKED EXAMPLE 1 *Working out the formula from reacting masses*

2.43 g of magnesium reacts with 0.20 g of hydrogen. What is the formula of magnesium hydride?

STEP 1 2.43 g of Mg $= \dfrac{2.43}{24.3} = 0.100$ moles **STEP 2** 0.20 g of H $= \dfrac{0.20}{1.0} = 0.20$ moles

STEP 3 Dividing through by the smallest number of moles (0.1 in this case) gives 1 Mg : 2 H. The empirical formula is therefore MgH_2.

WORKED EXAMPLE 2 *Working out the percentage by mass of an element*

Calculate the percentage by mass of nitrogen in ammonium sulfate, $(NH_4)_2SO_4$.

STEP 1 M_r $(NH_4)_2SO_4 = 2 \times (14.0 + 1.0 + 1.0 + 1.0 + 1.0) + 32.1 + (4 \times 16.0) = 132.1$

STEP 2 Mass of nitrogen in 1 mole of $(NH_4)_2SO_4 = 14.0 + 14.0 = 28.0$

STEP 3 % by mass $= \dfrac{\text{mass of element in 1 mole}}{M_r \text{ of compound}} \times 100 = \dfrac{28.0}{132.1} \times 100 = 21.2\%$

QUICK CHECK QUESTIONS

1 Work out the relative formula masses of:
   (a) MgO (b) Na_2SO_4 (c) $Pb(C_2H_5)_4$.
2 Give the empirical formulae of the following:
   (a) C_6H_6 (b) H_2O_2 (c) $C_{12}H_{22}O_{11}$.
3 How many moles of each substance are contained in:
   (a) 12.5 g of beryllium oxide
   (b) 22.0 g of CO_2?

4 Calculate the empirical formulae of compounds containing:
   (a) 3 g of magnesium and 2 g of oxygen only
   (b) 2 g of Mg, 1 g of C and 4 g of O only.
5 A compound contains 12.9% Be, 17.3% C and 69.8% O by mass. Calculate its empirical formula.
6 Calculate the percentage by mass of magnesium in $MgCO_3$.

A simple model of the atom

Chemical Ideas 2.1

Inside the atom

Mass number
= protons and neutrons ⟶ $^{23}_{11}\text{Na}$

Atomic number
= protons

^{23}Na

The atomic number can be omitted since all sodium atoms have an atomic number of 11.

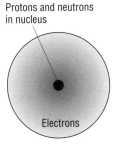

Protons and neutrons in nucleus

Electrons

Simple model of an atom.

The nuclear symbol tells you how many protons, electrons and neutrons there are in an atom of an element:

- number of protons = atomic number (bottom number)
- number of electrons in a neutral atom = atomic number (bottom number)
- number of neutrons = mass number – atomic number (top number – bottom number).

So a sodium atom has 11 protons, 11 electrons and 12 neutrons.

The three types of sub-atomic particles have different masses and charges.

Particle	Mass on relative atomic mass scale	Charge
proton	1	1+
neutron	1	0
electron	very small (0.000 55)	1–

Isotopes

Isotopes are atoms of the same element. They have the *same atomic number* (number of protons), but they have *different mass numbers* (different numbers of neutrons). For example, there are two common isotopes of chlorine:

- $^{35}_{17}\text{Cl}$ has 17 protons and 18 neutrons.
- $^{37}_{17}\text{Cl}$ has 17 protons and 20 neutrons.

Chlorine is a mixture of these isotopes. Knowing the **relative isotopic masses** and the **relative abundance** of the two isotopes allows you to calculate the relative atomic mass of the mixture. The relative atomic mass of the mixture is found by using the weighted average of the relative isotopic masses.

If you are asked to explain what the term 'isotopes' means using chlorine-35 and chlorine-37 then, after the definition, give the numbers of protons and neutrons in each isotope. Also, say what is the same and what is different between the isotopes.

If you are asked to quote the relative atomic mass to a certain number of decimal places, then do so.

WORKED EXAMPLE

If there is 75% of ^{35}Cl and 25% of ^{37}Cl in a sample of chlorine:

$$\text{average relative atomic mass} = \frac{(35 \times 75) + (37 \times 25)}{100} = 35.5$$

QUICK CHECK QUESTIONS

1. How many protons, neutrons and electrons are there in:
   (a) $^{3}_{1}\text{H}$ (b) $^{47}_{20}\text{Ca}$ (c) $^{23}_{11}\text{Na}$?
2. How many protons, neutrons and electrons are there in:
   (a) tellurium-122 (b) americium-241
   (c) carbon-13?

3. What are the missing numbers for these isotopes:
   (a) $^{13}_{?}\text{C}$ (b) $^{2}_{?}\text{H}$ (c) $^{4}_{?}\text{He}$?
4. Bromine has two isotopes – bromine-79 and bromine-81. The relative abundances are 50% for each isotope. Calculate the relative atomic mass to 1 decimal place.

Nuclear reactions

Chemical Ideas 2.2, Chemical Storylines EL1 and EL3

Radioactive isotopes

Some isotopes have **unstable nuclei** – this results in the isotope being **radioactive**.

Three types of emission can be produced spontaneously by radioactive nuclei. All three types of emission are capable of knocking electrons out of atoms, forming ions, so they are sometimes referred to as *ionising* radiation.

Radiation	What is it?	Relative charge	How does the nucleus change?	What can it be stopped by?	Deflection in electric field?
α	helium nuclei, $^{4}_{2}He$	+2	2 fewer protons; 2 fewer neutrons	paper or skin	low
β	electrons, $^{0}_{-1}e$	−1	1 more proton; 1 fewer neutron	aluminium foil	high
γ	electromagnetic radiation	none	no change	lead sheet	none

All three types of emission are dangerous to humans and can cause cancer.

Equations for radioactive decay

- α particle   e.g. $^{238}_{92}U \rightarrow {}^{234}_{90}Th + {}^{4}_{2}He$
- β particle   e.g. $^{14}_{6}C \rightarrow {}^{14}_{7}N + {}^{0}_{-1}e$
- γ radiation   there is no change in the atoms; just a release of energy from the nucleus. γ radiation often accompanies the emission of α and/or β particles.

The nuclear equations can be a little tricky so read exam questions carefully.

> With α and β emissions, the atomic number changes and a new element is formed. Look up the symbol for the new element in the Periodic Table.

> The sum of the mass numbers on the left of the arrow in the equation always adds up to the sum of the mass numbers on the right of the equation arrow. The same applies to the atomic numbers.

> an α particle is $^{4}_{2}He$ or $^{4}_{2}\alpha$
> a β particle is $^{0}_{-1}e$ or $^{0}_{-1}\beta$
> a neutron is $^{1}_{0}n$
> a proton is $^{1}_{1}p$

Example

Beryllium-9 can be bombarded with α particles. This causes neutrons to be released and carbon-12 is produced.

$$\underset{\substack{\text{look up the atomic number} \\ \text{in the Periodic Table}}}{^{9}_{4}Be} \quad + \quad \underset{\substack{\text{'bombarded} \\ \text{with'}}}{^{4}_{2}\alpha} \quad \rightarrow \quad \underset{\substack{\text{6 is the atomic} \\ \text{number for carbon}}}{^{12}_{6}C} \quad + \quad \underset{\text{'released'}}{^{1}_{0}n}$$

Half-life

The **half-life** is the time it takes for half the radioactive nuclei in a sample to decay. The half-life is fixed for any given radioisotope and is not affected by temperature.

Half-life calculations can be used to date archaeological artefacts made from living things – e.g. wood, cotton and bones. Ages of igneous rocks containing radioactive isotopes can also be estimated.

You can see examples of the calculations on page 5.

WORKED EXAMPLE 1   *How much?*

Iodine-131 has a half-life of 8 days. If you start with 24 g of the isotope, what mass remains after 32 days?

STEP 1 After 8 days, 12 g of the sample remains.

STEP 2 After 16 days, 6 g of the sample remains.

STEP 3 After 24 days, 3 g of the sample remains.

STEP 4 After 32 days, 1.5 g of the sample remains.

WORKED EXAMPLE 2   *How long?*

The half-life of carbon-14 is 5730 years. How long will it take for the sample's count rate to decrease from 160 counts min^{-1} to 20 counts min^{-1}?

STEP 1 After 1 half-life, the count rate would be 80 counts min^{-1}.

STEP 2 After 2 half-lives, the count rate would be 40 counts min^{-1}.

STEP 3 After 3 half-lives, the count rate would be 20 counts min^{-1}.

STEP 4 Time taken is 3 half-lives or 3 × 5730 years = 17 190 years.

Tracers

Tracers are radioactive isotopes whose decay is monitored. These isotopes can be used in medicine to aid diagnosis. The tracer is eaten, injected or drunk and then its pathway is followed using a Geiger counter. An isotope should have a half-life which is neither too short (or it will decay before tracing is complete) nor too long (or it will persist for too long in the body, potentially causing harm to the patient).

Radioisotopes can cause cancer, but tracers are relatively safe. Small doses are used to limit exposure. The benefits of early diagnosis are considered to outweigh the risks.

For more details about tracers, see Chemical Storylines EL3

Nuclear fusion

Nuclear fusion is the *joining* together of two or more nuclei to form a heavier nucleus of a new element. High temperatures and/or pressures are required to provide the energy needed to overcome the repulsion between two positive nuclei. Nuclear fusion occurs during star formation. Examples are:

For more details about nuclear fusion, see Chemical Storylines EL1

$$^1_1H + ^2_1H \rightarrow ^3_2He$$
$$3\,^4_2He \rightarrow ^{12}_6C$$

QUICK CHECK QUESTIONS

1 Complete these nuclear equations:
   (a) $^{47}_{20}Ca \rightarrow ^{0}_{-1}e +$ **(b)** $^{226}_{88}Ra \rightarrow ^{4}_{2}He +$
   (c) $^{241}_{95}Am \rightarrow ^{4}_{2}He +$ **(d)** $^{131}_{53}I \rightarrow ^{0}_{-1}e +$

2 An antique metal watch has hands painted with radioactive paint, which emits alpha radiation. Would this affect the wrist of the wearer?

3 $^{47}_{20}Ca$ has a half-life of 6.5 days. If you start with 2 g of the isotope, how much would be left after 26 days?

4 Name the equipment used to detect β particles.

5 The half-life of iodine-131 is 8 days. A sample of iodine has a count rate of 4000 count min^{-1}. How long would it take for the count rate to reach 125 counts min^{-1}?

Light and electrons

Chemical Ideas 6.1 and 2.3

Electrons always move to specific levels, never in between levels. Therefore, if you are asked to show an electron transition on a diagram always draw from one horizontal line to another – never start or finish between lines. For absorption spectra, the arrow goes up. For emission spectra, the arrow goes down.

Energy levels

An electron in a hydrogen atom (or any other atom) can occupy any one of the **fixed energy levels**. These different energy levels are the same for all hydrogen atoms. In the ground state, the electrons are closest to the nucleus and have the lowest energy. The difference in the energy levels (ΔE) *decreases* as the electron moves away from the nucleus.

Absorption spectra

An absorption spectrum seen on Earth is the spectrum of visible light (which looks like a rainbow) with black lines corresponding to the absorptions of energy by the electrons.

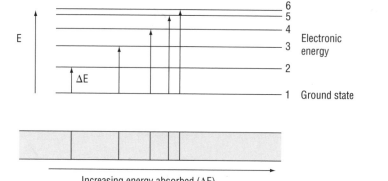

Increasing energy absorbed (ΔE)
Increasing frequency of light absorbed

An absorption spectrum.

These spectra are seen from Earth when atoms in the chromosphere around stars absorb light.

*   Electrons absorb a 'photon' or package of energy.

*   Excited electrons move up to a higher energy level – they are promoted. This is what produces the lines in an absorption spectrum.

*   The electromagnetic radiation absorbed by each of the hydrogen atoms has a definite frequency (ν) related to the difference in energy levels by $\Delta E = h\nu$.

Emission spectra

Similarities in the two types of spectra are:
*   the lines are in the same position
*   the lines become closer at higher frequency.

An emission spectrum has a black background with coloured lines on it. This is *different* to absorption spectra. These coloured lines correspond to the emissions of energy by the electrons.

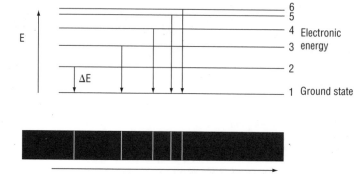

Increasing energy emitted (ΔE)
Increasing frequency of light emitted

An emission spectrum.

An emission spectrum is seen when a chemical burns with a coloured flame.

- Electrons first absorb a 'photon', or package of energy.
- Excited electrons move up to a higher energy level – they are promoted.
- Electrons then drop back to lower energy levels. This is what produces the lines in an emission spectrum.
- The electromagnetic radiation emitted by each of the hydrogen atoms has a definite frequency related to the difference in energy levels by $\Delta E = h\nu$. Since ΔE is different for each transition, the frequency (ν) is different for each transition, and hence so is the colour of the lines.

Electron shells

Energy levels are usually referred to as **electron shells** for atoms more complex than hydrogen. Each shell can hold a certain maximum number of electrons.

Electron shell	Maximum number of electrons
1st	2
2nd	8
3rd	18
4th	32

An electron will go into the lowest energy electron shell which is not fully occupied.

Using a Periodic Table, you should be able to work out the electron shell configurations for the first 36 elements. The atomic number (the lower number in the nuclear symbol from the Periodic Table) tells you how many electrons there are in a neutral atom.

Learn the electron shell configuration (electron arrangement) for the elements in the 4th Period (K–Kr). You should be able to work the electronic configuration for Periods 1–3.

The electron shell configuration can be used to identify the group and period to which an element belongs.

Magnesium (2.8.2) and calcium (2.8.8.2) have similar chemical properties because they both have two electrons in their outer shell. When they react, they both lose two electrons to become 2+ ions. Calcium is more reactive than magnesium because the outer shell electrons in calcium are further from the nucleus, are less firmly held, and so are more easily lost.

2 electrons in the outer shell, so Group 2

Mg 2.8.2

3 shells occupied so Period 3

Electron arrangements for Period 4 elements.

Element	Electron arrangement
K	2.8.8.1
Ca	2.8.8.2
Sc	2.8.9.2
Ti	2.8.10.2
V	2.8.11.2
Cr	2.8.13.1
Mn	2.8.13.2
Fe	2.8.14.2
Co	2.8.15.2
Ni	2.8.16.2
Cu	2.8.18.1
Zn	2.8.18.2
Ga	2.8.18.3
Ge	2.8.18.4
As	2.8.18.5
Se	2.8.18.6
Br	2.8.18.7
Kr	2.8.18.8

Note how Cr and Cu are unusual. You will learn about this in your later studies.

QUICK CHECK QUESTIONS

1 (a) Describe the appearance of an emission spectrum.
   (b) Why do hydrogen atoms give an emission spectrum, whereas hydrogen nuclei do not?
2 Draw a diagram of the energy levels in a hydrogen atom. Draw arrows on your diagram to show the origin of two of the lines in the hydrogen absorption spectrum.
3 Explain why hydrogen absorbs only certain frequencies of light.
4 Give the electron arrangement (in terms of electron shells) for carbon atoms and copper atoms.
5 Explain why strontium and calcium have similar chemical properties.
6 Explain why potassium is more reactive than sodium.

Chemical bonding and properties

Chemical Ideas 3.1

What type of bond?

The type of bond depends on the two atoms involved in the bond.

	Metal	Non-metal
Metal	metallic bonding	ionic bonding
Non-metal	ionic bonding	covalent bonding

Properties depend on bonding

Structure	Type of bonding	Melting point	Solubility in water	Electrical conductivity
ionic lattice e.g. NaCl	ionic	high	usually soluble	only if molten or in solution
giant covalent network e.g. SiO_2, diamond	covalent	high	insoluble	won't conduct, apart from graphite
simple molecular e.g. CO_2, H_2O	covalent	low	usually insoluble	won't conduct
metallic lattice e.g. Na, Mg	metallic	high	insoluble	will conduct

Ionic bonding

Atoms are usually more stable if they have a full outer shell of electrons. The metal atom *transfers* electron(s) to the non-metal atom so that all atoms end up with a full outer shell of electrons. This results in the formation of charged ions.

> Note that you only need to draw the outermost electrons.

> Be aware that dot–cross diagrams have their limitations. For example, in the ClO_3^- ion the central chlorine atom has expanded its outer shell to hold 12 electrons.

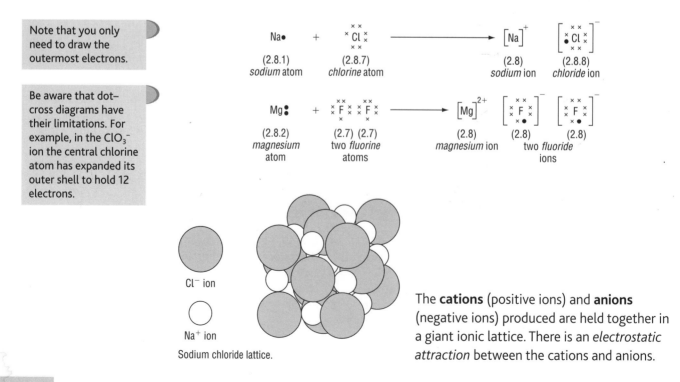

The **cations** (positive ions) and **anions** (negative ions) produced are held together in a giant ionic lattice. There is an *electrostatic attraction* between the cations and anions.

Sodium chloride lattice.

Covalent bonding

To attain a full outer shell of electrons, the two non-metal atoms involved in a covalent bond *share* a pair of electrons.

Molecular formula	Dot–cross diagram	Structural formula
H_2	H ⦁× H	H—H
NH_3	H ⦁ × H × N ⦁ × ⦁ H	H | H—N | H
H_2O	H ⦁ × H × O ⦁ ⦁ ⦁	H | H—O
O_2	×O× O××	O=O
N_2	×N× N××	N≡N
CO_2	×O× C ×O×	O=C=O

In some molecules, there is a special type of covalent bond called a **dative covalent bond**. Both of the electrons in a dative covalent bond come from the same atom.

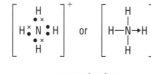

×C ⦁ O×   or   C≡O

Dot–cross   An arrow is
diagram   sometimes used to
show a dative bond

carbon monoxide.

$$\left[\begin{array}{c} H \\ × ⦁ \\ H × N ⦁ H \\ ⦁ × \\ H \end{array} \right]^+ \quad \text{or} \quad \left[\begin{array}{c} H \\ | \\ H—N{\rightarrow}H \\ | \\ H \end{array} \right]^+$$

ammonium ion.

> The atoms involved in the covalent bond are held together by an **electrostatic attraction** between the positive nuclei of the two atoms and the shared pair of negative electrons.

Metallic bonding

The metal ions are arranged regularly in a lattice. The outer shell electrons are shared by all the ions and are said to be **delocalised**. The 'sea' of electrons are free to move and metals conduct electricity because of this.

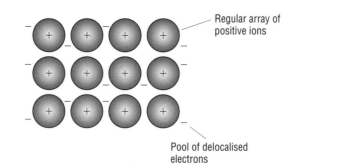

Regular array of positive ions

Pool of delocalised electrons

> The metallic bond that holds the particles in a metal together is the *electrostatic attraction* between the positive metal ions and the negative electrons.

QUICK CHECK QUESTIONS

1 Describe the type of bonding in:
    (a) CH_4 (b) KF   (c) SiH_4
    (d) Mg   (e) Hg
2 Draw dot–cross diagrams for:
    (a) CH_4 (b) C_2H_6 (c) CH_3CH_2OH
    (d) CO_2 (e) CH_2CH_2 (f) CH_3OCH_3
3 Draw dot–cross diagrams for:
    (a) LiBr   (b) Na_2O (c) CaO
    (d) $CaCl_2$ (e) Al_2O_3

4 What type of *structure* do the following have:
    (a) magnesium chloride
    (b) water
    (c) graphite
    (d) oxygen
    (e) magnesium
    (f) sodium bromide?
5 Which would you expect to be the most soluble in water: oxygen or magnesium chloride?

Shapes of molecules
Chemical Ideas 3.2

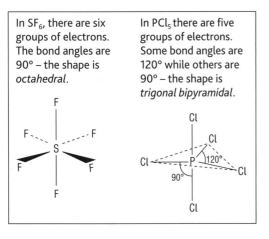

In SF_6, there are six groups of electrons. The bond angles are 90° – the shape is *octahedral*.

In PCl_5 there are five groups of electrons. Some bond angles are 120° while others are 90° – the shape is *trigonal bipyramidal*.

- The shape of a molecule depends on the number of *groups* of electrons.
- A group of electrons could be a bonding pair (a single bond), two bonding pairs (a double bond), three bonding pairs (a triple bond) or a lone pair.
- Groups of electrons repel each other.
- Groups of electrons will arrange themselves so as *to be as far apart in space as possible.*

Four groups of electrons results in bond angles of 109°.

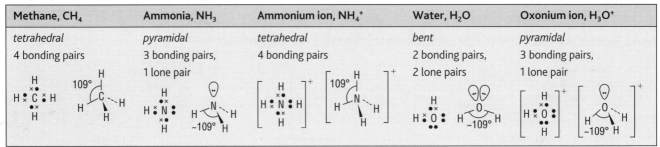

Methane, CH_4	Ammonia, NH_3	Ammonium ion, NH_4^+	Water, H_2O	Oxonium ion, H_3O^+
tetrahedral	*pyramidal*	*tetrahedral*	*bent*	*pyramidal*
4 bonding pairs	3 bonding pairs, 1 lone pair	4 bonding pairs	2 bonding pairs, 2 lone pairs	3 bonding pairs, 1 lone pair

Three groups of electrons results in bond angles of 120°.

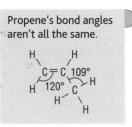

Propene's bond angles aren't all the same.

Boron trifluoride, BF_3	Ethene, C_2H_4	Methanal, CH_2O
planar triangular	*planar triangular*	*planar triangular*
3 bonding pairs	2 bonding pairs, 1 double bond for each C	2 bonding pairs, 1 double bond

Two groups of electrons results in bond angles of 180°.

Beryllium chloride, $BeCl_2$	Ethyne, C_2H_2	Carbon dioxide, CO_2
linear	*linear*	*linear*
2 bonding pairs	1 triple bond, 1 bonding pair for each C	2 double bonds

QUICK CHECK QUESTIONS

1 What are the bond angles in:
   (a) SiH_4
   (b) PH_3
   (c) H_2S?

2 What are the bond angles **a** to **f** in the following molecules?

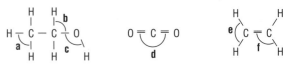

Periodicity and the Periodic Table

Chemical Ideas 11.1 and Chemical Storylines EL4

Mendeleev and the Periodic Table

- Mendeleev arranged the known elements in order of *relative atomic mass*.
- Elements with similar physical and chemical properties were in the same group. An example of this is lithium, sodium and potassium, which are reactive metals with low melting points.
- He then swapped elements over if he thought that they fitted better into another group based on their physical and chemical properties. An example of this is iodine and tellurium. Although the relative atomic mass of tellurium is higher than iodine, their properties were not similar to other elements in the same group, unless they were swapped over.
- He left *gaps* for elements which he thought were yet to be discovered.
- He made predictions about the properties of elements yet to be discovered. Later, his predictions were found to be very close to the actual properties. This validated his version of the Periodic Table in the opinion of other chemists.

We now arrange the elements in order of *atomic number* (sometimes called proton number) rather than relative atomic mass.

> Mendeleev arranged his elements in order of relative atomic mass, not mass number or atomic number. Some candidates lose marks by using incorrect terminology.

Periodicity

> See pages 8 and 9 for more about structures of elements.

Periodicity is exhibited when:
- there is a *regular pattern* in a property as you go across a period
- the regular pattern is *repeated* in other periods.

Example of periodicity: melting point and boiling point

The pattern is for the melting point to increase and then decrease across a period. This pattern is repeated in more than one period. There is periodicity in melting points. Similarly there is periodicity in boiling points.

Melting point increases and then decreases across Period 2 (atomic numbers 3–10). There is the same increase and decrease in Period 3 (atomic numbers 11–18).

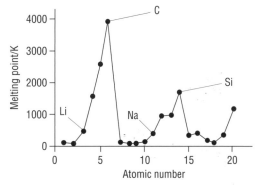

> Metals are on the left-hand side of a period and have metallic lattice structures. Some Group 4 elements have covalent network structures and so tend to have higher melting and boiling points. Other non-metals found on the right-hand side of a period have simple molecular structures and so tend to have lower melting and boiling points.

QUICK CHECK QUESTIONS

1 By which property do we now arrange the elements in the Periodic Table?

2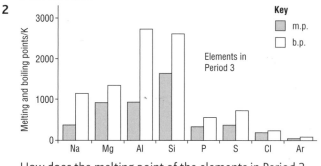

How does the melting point of the elements in Period 3 change on going from left to right?

3 The relative atomic mass of iodine (126.9) is smaller than that of tellurium (127.6).
   (a) Why did Mendeleev swap them over so that tellurium was in Group 6 and iodine was in Group 7?
   (b) Predict the melting point of fluorine.

Element	Melting point (°C)
F	?
Cl	−101
Br	7
I	114

Group 2 and writing balanced equations

Chemical Ideas 1.2, 3.1 and 11.2

Writing the formulae you will need

There are some formulae which it would be useful to learn:

water	H_2O
carbon dioxide	CO_2
hydrogen	H_2
hydrochloric acid	HCl
sulfuric acid	H_2SO_4

There are some ion formulae which you should also learn:

beryllium	Be^{2+}	chloride	Cl^-
magnesium	Mg^{2+}	hydroxide	OH^-
calcium	Ca^{2+}	oxide	O^{2-}
strontium	Sr^{2+}	carbonate	CO_3^{2-}
barium	Ba^{2+}	sulfate	SO_4^{2-}
sodium	Na^+	nitrate(V)	NO_3^-
potassium	K^+	ammonium	NH_4^+
		hydrogen carbonate	HCO_3^-

WORKED EXAMPLE 1

To work out the formula of magnesium oxide:

$1 \times Mg^{2+}$ is balanced by $1 \times O^{2-}$

Formula is MgO

Remember – formulae don't have the number '1' in them, so it is MgO rather than Mg_1O_1.

WORKED EXAMPLE 2

To work out the formula of magnesium chloride:

$1 \times Mg^{2+}$ is balanced by $2 \times Cl^-$

Formula is $MgCl_2$

Remember – if the formula of an ionic compound is correct the *total* of the charges on all the positive and negative ions will be equal to zero.

WORKED EXAMPLE 3

To work out the formula of magnesium hydroxide:

$1 \times Mg^{2+}$ is balanced by $2 \times OH^-$

Formula is $Mg(OH)_2$

Remember – brackets are used if there are two or more of a complex ion such as OH^-, CO_3^{2-} and SO_4^{2-}.

Balancing equations

WORKED EXAMPLE 4

STEP 1 Write the word equation:

calcium + water → calcium hydroxide + hydrogen

STEP 2 Put in the formulae:

$Ca + H_2O \rightarrow Ca(OH)_2 + H_2$

STEP 3 Balance the equation using large numbers before the formulae:

$Ca + 2H_2O \rightarrow Ca(OH)_2 + H_2$

Never change the formulae (i.e. do not change the subscript values in an equation).

Check	Left	Right
Ca	1	1
O	2	2
H	4	4

STEP 4 Add the state symbols if asked for in the exam

$Ca(s) + 2H_2O(l) \rightarrow Ca(OH)_2(aq) + H_2(g)$

Remember – water is a liquid and (l) is used; but for solutions (aq) is used.

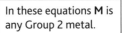

Reactions of the elements in Group 2

The Group 2 elements are Be, Mg, Ca, Sr, Ba and Ra.

- The metals react with water to give the metal hydroxide and hydrogen.

metal + water → metal hydroxide + hydrogen

$M(s) + 2H_2O(l) \rightarrow M(OH)_2(s) + H_2(g)$ ⟵

As is typical for metals, the reactions become more vigorous as you go down the group.

> In these equations **M** is any Group 2 metal.

> Group 2 elements have similar chemical reactions as they all have two electrons in their outer shell.

The oxides

- The oxides react with water to produce an alkaline solution of the hydroxide.

metal oxide + water → metal hydroxide

$MO(s) + H_2O(l) \rightarrow M(OH)_2(s)$ ⟵

- The oxides react with acids and so act as bases.

metal oxide + acid → salt + water

$MO(s) + H_2SO_4(aq) \rightarrow MSO_4(aq) + H_2O(l)$

> In these equations, the state of $M(OH)_2$ could be (s) or (aq) depending on the solubility of the metal hydroxide.

The hydroxides

The hydroxides become more soluble as you go down the group. The solutions produced are alkaline since the solutions contain $OH^-(aq)$ (pH > 7).

The hydroxides react with acids to produce a salt and water.

metal hydroxide + acid → salt + water

$M(OH)_2(s) + 2HCl(aq) \rightarrow MCl_2(aq) + 2H_2O(l)$ ⟵

The carbonates

The carbonates become less soluble as you go down the group.

The carbonates undergo thermal decomposition on heating to give the metal oxide and carbon dioxide.

metal carbonate → metal oxide + carbon dioxide

$MCO_3(s) \rightarrow MO(s) + CO_2(g)$

Thermal stability increases down the group. This means that barium carbonate is less likely to break down on heating than is magnesium carbonate.

QUICK CHECK QUESTIONS

1 Write a balanced equation with state symbols for the reaction of strontium with water.
2 Write the formulae for:
   (a) calcium chloride **(b)** strontium hydroxide
   (c) calcium sulfate **(d)** barium hydroxide.
3 Strontium hydroxide solution contains hydroxide ions. How would you test the solution for the presence of hydroxide ions and what would be the result?
4 Write a balanced equation, with state symbols, for the thermal decomposition of calcium carbonate.
5 Which one in each pair is the most soluble?
   (a) $Mg(OH)_2$ or $Ba(OH)_2$ **(b)** $MgCO_3$ or $BaCO_3$
6 Write a balanced equation, with state symbols, for the reaction of magnesium oxide with hydrochloric acid.
7 Write a balanced equation, with state symbols, for the reaction between magnesium hydroxide and sulfuric acid.

The mass spectrometer

Chemical Ideas 6.5

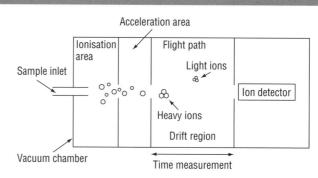

A schematic diagram showing the main parts of a time-of-flight mass spectrometer.

You may be asked to label a diagram of a mass spectrometer, or to describe and explain what happens in each part of the mass spectrometer.

You need to know what happens in each part of the mass spectrometer.

- *Sample inlet* – gases or liquids are simply injected but solids are heated to vaporise them.
- *Ionisation area* – a heated filament produces high-energy electrons. These electrons bombard any atoms or molecules in the sample and knock electrons out. Cations (positive ions) are formed: $X(g) + e^- \rightarrow X^+(g) + 2e^-$.
- *Acceleration area* – an electric field is used to accelerate any ions so that they all have the same kinetic energy.
- *Drift region* – there is a vacuum here so that ions do not collide with air molecules, which could change the direction of their flight path. Since kinetic energy = mass × velocity^2 and all ions have the same kinetic energy, heavier ions move through this region more slowly than light ions.
- *Ion detector* – light ions reach the detector before heavier ones. A computer system converts the information into a mass spectrum. Only positive ions are detected after fragmentation.

The mass spectrometer and isotopes

You may be asked to calculate the relative atomic mass of an element from its mass spectrum.

You may be asked to sketch a mass spectrum using data from a mass spectrometer.

The x-axis shows the mass to charge ratio (*m/z*). Assume that all the ions formed have a 1+ charge. This means *m/z* is the *same* as the mass of the ion detected.

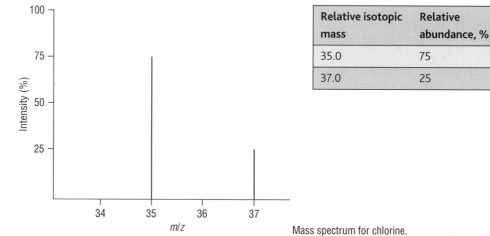

Relative isotopic mass	Relative abundance, %
35.0	75
37.0	25

Mass spectrum for chlorine.

If there is 75% of ^{35}Cl and 25% of ^{37}Cl:

$$\text{relative atomic mass} = \text{average mass of 100 atoms} = \frac{(35.0 \times 75) + (37.0 \times 25)}{100} = 35.5$$

The mass spectrometer and molecules

The mass spectrum for a compound can be much more complex than the mass spectrum for an element.

The ion with the greatest mass corresponds to the molecular mass of the sample compound. This is called the **molecular ion**, and corresponds to the parent molecule minus an electron. In this case $m/z = 46$. The most intense peak is called the **base peak**. There are many peaks because **fragments** are formed in the ionisation chamber.

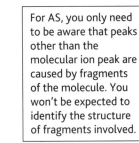

Mass spectrum of ethanol, CH_3CH_2OH.

The **molecular ion** breaks down into fragments. If the fragment has a positive charge then it will be accelerated by the electrical field and detected later in the mass spectrometer.

You can see in this flow diagram how the peaks at $m/z = 46, 45, 29$ and 31 have formed.

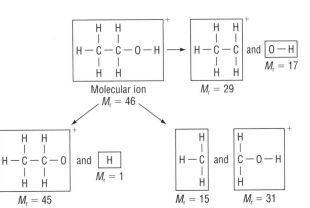

For AS, you only need to be aware that peaks other than the molecular ion peak are caused by fragments of the molecule. You won't be expected to identify the structure of fragments involved.

QUICK CHECK QUESTIONS

1   Calculate the relative atomic mass of magnesium from this data. Give your answer to 1 decimal place.

Relative atomic mass of isotope	Relative abundance (%)
24.0	70
25.0	19
26.0	11

2   Sketch the mass spectrum produced by the sample of magnesium in question 1.

3   When a time-of-flight mass spectrometer is used, how are the ions accelerated?

4   When a time-of-flight mass spectrometer is used, why do the ions formed take different times to pass through the drift region?

5   The mass spectrum above right is that of ethanoic acid (present in vinegar).
   (a)   Why are there so many peaks?
   (b)   Which fragment is responsible for the peak at 60?
   (c)   Which fragment is responsible for the peak at 15?
   (d)   Which fragment is responsible for the peak at 45?
   (e)   What name to we give to the peak at 43?

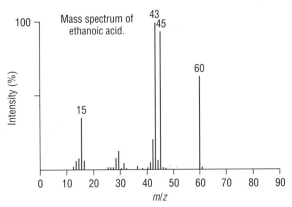

Mass spectrum of ethanoic acid.

Structural formula of ethanoic acid.

6   The mass spectrum below is that of butane.
   (a)   What is the M_r of butane?
   (b)   Which ion is responsible for the peak at 15?
   (c)   Which ion is responsible for the peak at 29?

Structural formula of butane.

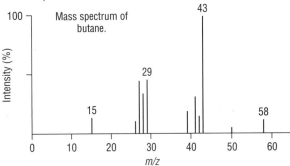

Mass spectrum of butane.

Developing Fuels (DF)

This module tells the story of petrol: what it is and how it is made. It describes the work of chemists in improving fuels for motor vehicles and in searching for alternative fuels for the future. As you work through this module you will cover the following concepts. 'CI' refers to sections in your *Chemical Ideas* textbook. 'Storylines' refers to your *Chemical Storylines* textbook.

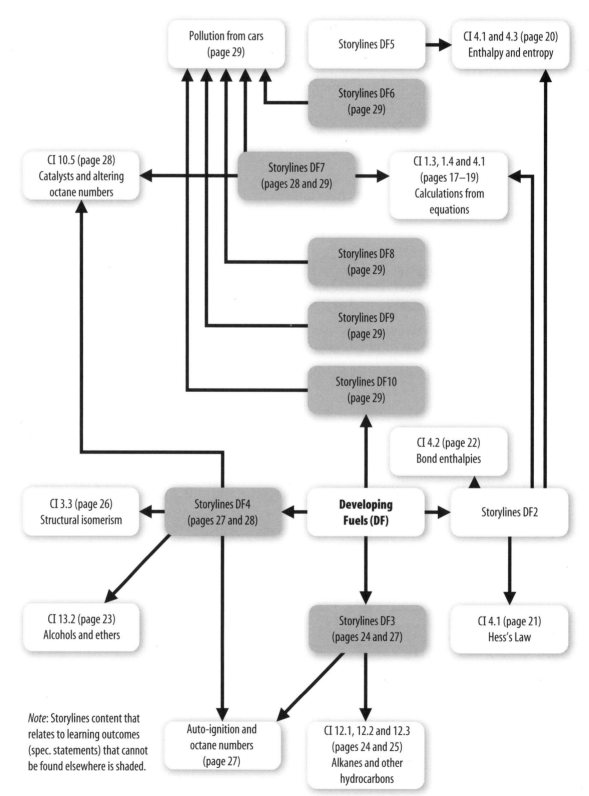

Note: Storylines content that relates to learning outcomes (spec. statements) that cannot be found elsewhere is shaded.

Calculations from equations

Chemical Ideas 1.3, 1.4 and 4.1

Remember – the large numbers *in front* of formulae in an equation tell you the number of moles involved.

$$CH_4 + 2O_2 \rightarrow CO_2 + 2H_2O$$

1 mole of methane reacts with *2 moles* of oxygen to produce *1 mole* of carbon dioxide and *2 moles* of water.

Working out masses

To work out the mass of 1 mole of a compound (M_r) add up the relative atomic masses (A_r) of the atoms in the compound.

e.g. $CaCO_3$ = 40.1 + 12.0 + 16.0 + 16.0 + 16.0 = 100.1 (remember, no units)

WORKED EXAMPLE

What mass of magnesium oxide is produced when 1.2 g of magnesium reacts with carbon dioxide? (A_r: Mg, 24.3; O, 16.0)

STEP 1 Underline the substances whose:
- mass you are given
- mass you want to find.

$\underline{2Mg} \quad + CO_2 \rightarrow \quad \underline{2MgO} + C$

STEP 2 Indicate the number of moles involved:

2 moles   2 moles

STEP 3 Calculate the masses:

48.6 g   80.6 g

STEP 4 Convert to the mass given in the question:

$\dfrac{48.6}{48.6} \times 1.2 = 1.2\,g$

To work out the mass of 2 moles of a compound, multiply the M_r by 2 because

mass (g) = amount (moles) × M_r

STEP 5 Convert the other mass in the same way:

$\dfrac{80.6}{48.6} \times 1.2 = 2.0\,g$

STEP 6 Write down the answer:

2.0 g of MgO is produced when 1.2 g of Mg reacts with CO_2.

Note: In **STEP 4** – converting to the mass given in the question – the A_r of magnesium is 24.3; there are 2 moles in this equation; 24.3 × 2 = 48.6. Dividing by 48.6 gives 1.0 g of magnesium; then multiplying by 1.2 gives 1.2 g, as in the question.

Note: In **STEP 5** – converting the other mass in the same way – converting the other mass in the same way keeps the proportions correct.

The approach shown in Steps 4 and 5 has been used in all the worked examples.

Working out volumes of gases

The volume of 1 mole of gas will be given in exam questions.

The volume of 1 mole of a gaseous substance is $24\,dm^3$ at room temperature and pressure (25 °C and 1 atmosphere pressure).

WORKED EXAMPLE

What volume of carbon dioxide is produced in the complete combustion of $6.0\,dm^3$ of methane?

STEP 1 Underline the substances whose:
- volume you are given
- volume you want to find.

$$\underline{CH_4} + 2O_2 \rightarrow \underline{CO_2} + 2H_2O$$

STEP 2 Indicate the number of moles involved: 1 mole   1 mole

STEP 3 Calculate the volumes of this many moles: $24\,dm^3$ $24\,dm^3$

If you need to work out the volume of 2 moles of a gas under standard conditions, multiply 24 by 2.

STEP 4 Convert to the volume given in the question: $\dfrac{24}{24} \times 6.0 = 6.0\,dm^3$

STEP 5 Convert the other volume in the same way: $\dfrac{24}{24} \times 6.0 = 6.0\,dm^3$

STEP 6 Write down the answer: $6.0\,dm^3$ of CO_2 is produced in the complete combustion of $6.0\,dm^3$ of methane.

Working with both mass and volume

WORKED EXAMPLE

What volume of carbon dioxide is produced when 20 g of calcium carbonate is heated?

STEP 1 Underline the substances whose:
- mass or volume you are given
- mass or volume you want to find.

$$\underline{CaCO_3} \rightarrow CaO + \underline{CO_2}$$

STEP 2 Indicate the number of moles involved: 1 mole   1 mole

STEP 3 Calculate the mass or volume: 100.1 g $24\,dm^3$

STEP 4 Convert to the mass or volume given: $\dfrac{100.1}{100.1} \times 20 = 20\,g$

STEP 5 Convert the other mass or volume in the same way: $\dfrac{24}{100.1} \times 20 = 4.8\,dm^3$

STEP 6 Write down the answer: $4.8\,dm^3$ of CO_2 is produced when 20 g of calcium carbonate is heated.

Working with enthalpy changes

The **enthalpy change of combustion** tells you the amount of energy released when *1 mole* of a fuel is burnt completely.

> To work out energy released when 2 moles of a fuel are burnt, multiply the enthalpy change of combustion by 2.

WORKED EXAMPLE

The enthalpy of combustion for methane is $-890\,kJ\,mol^{-1}$. Calculate the energy released when 3.20 g of methane burn completely.

STEP 1 Underline:
- the substances whose masses or volumes you are given
- the enthalpy change you are given.

$$\underline{CH_4} + 2O_2 \rightarrow CO_2 + 2H_2O \qquad \Delta H = \underline{-890}\,kJ\,mol^{-1}$$

STEP 2 Indicate the number of moles involved: 1 mole $\Delta H = -890\,kJ\,mol^{-1}$

STEP 3 Calculate the mass of this many moles: 16.0 g

STEP 4 Convert to the mass or volume given: $\dfrac{16.0}{16.0} \times 3.20 = 3.20\,g$

STEP 5 Convert the energy released in the same way:

$$\dfrac{-890}{16.0} \times 3.20$$
$$= -178\,kJ\,mol^{-1}$$

STEP 6 Write down the answer: 178 kJ of energy is released when 3.20 g of methane burns completely.

Table of relative atomic masses, A_r.

H	1.0
C	12.0
N	14.0
O	16.0
Na	23.0
Mg	24.3
Ca	40.1
Fe	55.8

QUICK CHECK QUESTIONS

Assume that 24 dm^3 is the volume of 1.0 moles of any gas under the given conditions.

1 What mass of iron(III) oxide is produced when 11.2 g of iron burns?
$$4Fe(s) + 3O_2(g) \rightarrow 2Fe_2O_3(s)$$

2 Calculate the volume of hydrogen that would react exactly with 6 dm^3 of carbon monoxide.
$$6CO(g) + 13H_2(g) \rightarrow C_6H_{14}(l) + 6H_2O(l)$$

3 Calculate the volume of nitrogen gas that would be produced by the thermal decomposition of 0.65 g of sodium azide.
$$2NaN_3(s) \rightarrow 2Na(s) + 3N_2(g)$$

4 The enthalpy change of combustion of octane is $-5470\,kJ\,mol^{-1}$. How much energy is released when 5.7 g of octane burns completely?
$$C_8H_{18}(l) + 12\tfrac{1}{2}O_2(g) \rightarrow 8CO_2(g) + 9H_2O(l)$$

5 Calculate the mass of hexane, C_6H_{14}, that would be formed from 6.0 dm^3 of carbon monoxide, CO.
$$6CO(g) + 13H_2(g) \rightarrow C_6H_{14}(l) + 6H_2O(l)$$

6 Calculate the energy released (in kJ) by burning 100 g of hydrogen completely. The enthalpy of combustion of hydrogen $[H_2(g) + \tfrac{1}{2}O_2(g) \rightarrow H_2O(l)]$ is $-243\,kJ\,mol^{-1}$.

Enthalpy and entropy

Chemical Ideas 4.1 and 4.3

Enthalpy

Standard conditions are 1 atmosphere pressure and 298 K (25 °C).

- An **exothermic** reaction *gives out* energy from the system to the surroundings. The temperature of the surroundings increases, ΔH is *negative*.
- An **endothermic** reaction *takes energy into* the system from the surroundings. The temperature of the surroundings decreases, ΔH is *positive*.

If the equation is doubled, then so is ΔH.

- The **standard enthalpy change of combustion**, $\Delta H_c^{\ominus}$, is the enthalpy change when *1 mole* of a substance *burns completely* in oxygen under *standard conditions*.

$$CH_4(g) + 2O_2(g) \rightarrow CO_2(g) + 2H_2O(l); \Delta H_c^{\ominus} = -890 \, kJ \, mol^{-1}$$

- The **standard enthalpy change of formation**, $\Delta H_f^{\ominus}$, is the enthalpy change when *1 mole* of a substance is *formed* from its *constituent elements*. Both the reactants and products are in their *standard state*.

Remember to put H_2 or O_2 in equations for enthalpy of formation.

$$H_2(g) + \tfrac{1}{2}O_2(g) \rightarrow H_2O(l); \Delta H_f^{\ominus} = -286 \, kJ \, mol^{-1}$$

- **Other reactions** – the enthalpy change for other reactions under standard conditions may be given the symbol $\Delta H_r^{\ominus}$

A temperature rise in K is the same as in °C.

Measuring $\Delta H_c^{\ominus}$ in the laboratory

m = mass of *water* (g)
c = specific heat capacity of water ($4.18 \, J \, g^{-1} \, K^{-1}$)
ΔT = change in temperature (K)

Record the temperature rise when a known volume of water is heated by the complete combustion of a measured mass of fuel.

The energy transferred = $m \times c \times \Delta T$

See **Activity DF2.1** for details. The value obtained will be much lower than the data sheet value due to the heat loss to the surroundings.

You can now calculate the enthalpy change for the combustion of 1 mole of the fuel used.

Thermometer — Cover, to reduce heat loss

Metal calorimeter

Clamp

Draught shield — Water

Spirit burner

Liquid fuel

Simple apparatus for measuring $\Delta H_c^{\ominus}$.

Entropy

Don't use the word 'atom' in answers to entropy questions.

- Entropy is a measure of the number of ways in which *particles* can be arranged.
- Gases have greater entropy than liquids; liquids have greater entropy than solids.
- Mixtures (e.g. solutions) have a greater entropy than the unmixed constituents.
- If the number of *particles* increases during the course of a reaction then entropy usually increases.

QUICK CHECK QUESTIONS

1 Draw a labelled diagram of the apparatus you would use to measure the enthalpy change of combustion of propanol.

2 The temperature rise when excess zinc was added to copper sulfate solution was found to be 10 °C.
Heat transferred = 25 × 4.18 × 10 J.

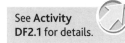

See **Activity DF2.1** for details.

(a) What two assumptions did they make in this calculation?

(b) If the concentration of the copper sulfate solution was 0.1 mol dm⁻³ what would be the enthalpy change for this reaction ($\Delta H_r^{\ominus}$)?

3 When this reaction happens, would the entropy increase, decrease or stay the same? Explain your answer.

$$2NaN_3(s) \rightarrow 2Na(l) + 3N_2(g)$$

4 Would you expect the entropy of the system to change when liquefied propane changes into gas? Explain why.

Hess's law

Chemical Ideas 4.1

Hess's law

Hess's law states that as long as the starting and finishing points are the same, the enthalpy change for a chemical reaction will always be the same, no matter how you go from start to finish. It is useful for calculating unknown enthalpy changes from ones for which data is available.

Enthalpy cycles

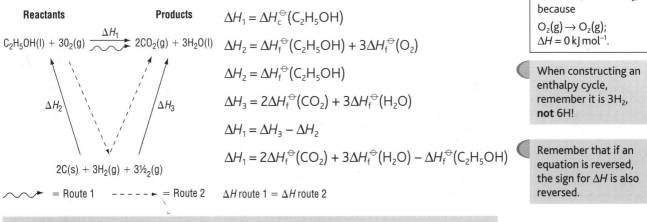

Reactants   **Products**

$C_2H_5OH(l) + 3O_2(g) \xrightarrow{\Delta H_1} 2CO_2(g) + 3H_2O(l)$

ΔH_2 ΔH_3

$2C(s) + 3H_2(g) + 3\frac{1}{2}O_2(g)$

$\sim\!\!\sim\!\!\rightarrow$ = Route 1 $----\blacktriangleright$ = Route 2 ΔH route 1 = ΔH route 2

$\Delta H_1 = \Delta H_c^{\ominus}(C_2H_5OH)$

$\Delta H_2 = \Delta H_f^{\ominus}(C_2H_5OH) + 3\Delta H_f^{\ominus}(O_2)$

$\Delta H_2 = \Delta H_f^{\ominus}(C_2H_5OH)$

$\Delta H_3 = 2\Delta H_f^{\ominus}(CO_2) + 3\Delta H_f^{\ominus}(H_2O)$

$\Delta H_1 = \Delta H_3 - \Delta H_2$

$\Delta H_1 = 2\Delta H_f^{\ominus}(CO_2) + 3\Delta H_f^{\ominus}(H_2O) - \Delta H_f^{\ominus}(C_2H_5OH)$

> ΔH_f is zero for elements, such as O_2, because
> $O_2(g) \rightarrow O_2(g)$;
> $\Delta H = 0 \, kJ \, mol^{-1}$.

> When constructing an enthalpy cycle, remember it is $3H_2$, **not** $6H$!

> Remember that if an equation is reversed, the sign for ΔH is also reversed.

WORKED EXAMPLE

Calculate the standard enthalpy change of combustion for ethanol using the standard enthalpy changes of formation on the right.

Compound	$\Delta H_f^{\ominus}$ (kJ mol^{-1})
$C_2H_5OH(l)$	−277
$CO_2(g)$	−394
$H_2O(l)$	−286

$C_2H_5OH(l) + 3O_2(g) \rightarrow 2CO_2(g) + 3H_2O(l)$

$\Delta H_c^{\ominus} = \Delta H_f(products) - \Delta H_f^{\ominus}(reactants)$

$\Delta H_f^{\ominus}(products) = (2 \times -394) + (3 \times -286) = -1646$

$\Delta H_f^{\ominus}(reactants) = -277$

$\Delta H_c^{\ominus} = -1646 - (-277) = -1369 \, kJ \, mol^{-1}$

The sign and the units may carry a mark. Don't forget them. If the sign is +, then put the +, don't forget it.

> $\Delta H_1 = \Delta H_f(products) - \Delta H_f(reactants)$ is useful when calculating an enthalpy of combustion from enthalpies of formation *only*.

> Remember to multiply $\Delta H_f^{\ominus}$ by the number of moles in the equation.

QUICK CHECK QUESTIONS

1. Draw and label an enthalpy cycle involving $\Delta H_c^{\ominus}(N_2H_4)$, $\Delta H_f^{\ominus}(N_2H_4)$ and $\Delta H_f^{\ominus}(H_2O)$.
   $$N_2H_4(l) + O_2(g) \rightarrow N_2(g) + 2H_2O(l)$$

2. Use the data to calculate $\Delta H_c^{\ominus}$ for carbon monoxide.
   $\Delta H_f^{\ominus}(CO) = -110.5 \, kJ \, mol^{-1}$; $\Delta H_f^{\ominus}(CO_2) = -393.5 \, kJ \, mol^{-1}$
   $$CO(g) + \frac{1}{2}O_2(g) \rightarrow CO_2(g)$$

3. Use the data to calculate $\Delta H_c^{\ominus}$ for hydrazine.
   $\Delta H_f^{\ominus}(N_2H_4) = +51 \, kJ \, mol^{-1}$; $\Delta H_f^{\ominus}(H_2O) = -286 \, kJ \, mol^{-1}$
   $$N_2H_4(l) + O_2(g) \rightarrow N_2(g) + 2H_2O(l)$$

4. **(a)** What is the relationship between ΔH_1, ΔH_2 and ΔH_3?

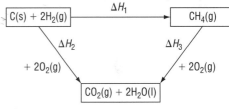

$C(s) + 2H_2(g) \xrightarrow{\Delta H_1} CH_4(g)$

ΔH_2 ΔH_3

$+ 2O_2(g)$ $+ 2O_2(g)$

$CO_2(g) + 2H_2O(l)$

   (b) Calculate ΔH_1.
   $\Delta H_c^{\ominus}(C) = -394 \, kJ \, mol^{-1}$; $\Delta H_c^{\ominus}(H_2) = -286 \, kJ \, mol^{-1}$;
   $\Delta H_c^{\ominus}(CH_4) = -890 \, kJ \, mol^{-1}$

Bond enthalpies

Chemical Ideas 4.2

Bond	Bond enthalpy (kJ mol^{-1})
C–C	+347
C–H	+413
O=O	+498
O–H	+464
C=O (in CO_2)	+805
C–O	+358
H–H	+436
C=C	+612

It is a good idea to draw each molecule in the equation and mark the bonds as you count them!

$$\begin{array}{ccc} & H & & H \\ & + & & + \\ H + & C & + & C & + H \\ & + & & + \\ & H & & H \end{array}$$

A common mistake is to count two C–C bonds because there are two carbon atoms.

Make sure that you know what O_2, CO_2 and H_2O look like.

Bond enthalpy and enthalpy cycles

Bond enthalpy is the *average* energy required to *break* the bonds in 1 mole of *gaseous* compounds.

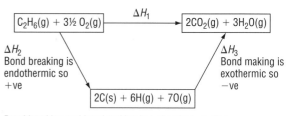

Bond breaking and bond making in a chemical reaction.

$$\Delta H_1 = \Delta H_2 + \Delta H_3$$

Bond enthalpy calculations

WORKED EXAMPLE

Calculate the enthalpy change of combustion for ethane (C_2H_6) using the bond enthalpy data on the left.

$$C_2H_6 + 3\tfrac{1}{2}O_2 \rightarrow 2CO_2 + 3H_2O$$

$$C_2H_6 \quad + \quad 3\tfrac{1}{2}\,O_2 \quad \rightarrow \quad 2CO_2 \quad + \quad 3H_2O$$

STEP 1 Bonds broken

1 × C–C (347)	= +347
6 × C–H (413)	= +2478
3.5 × O=O (498)	= +1743
Total	= **+4568**

(+ sign because bond breaking)

STEP 2 Bonds made

4 × C=O (805)	= –3220
6 × O–H (464)	= –2784
Total	= **–6004**

(– sign because bond making)

STEP 3 $\Delta H_c^{\ominus}(C_2H_6) = +4568 - 6004 = -1436\,\text{kJ mol}^{-1}$

This value is not exactly the same as quoted values, because:

- a bond enthalpy is an average energy needed to break that particular bond and is not specific to the molecule in an equation
- bond enthalpies are for gaseous molecules and this may not be their standard state (the state at 298 K and 1 atmosphere).

ΔH_c increases by a regular amount with the addition of each –CH_2–. The same additional bonds are broken and made.

Bond strengths

- The greater the bond enthalpy, the stronger the bond.
- Short bonds are stronger than long bonds.
- C=C is shorter and stronger than C–C.

QUICK CHECK QUESTIONS

1 Use bond enthalpy data to calculate a value for the enthalpy of combustion of hydrogen.

$$H_2 + \tfrac{1}{2}O_2 \rightarrow H_2O$$

2 Draw the enthalpy cycle (using the one at the top of this page as a model) to show the bond breaking and bond making in the complete combustion of propane.

3 Use bond enthalpy data to calculate the enthalpy of combustion of hept-1-ene.

$$C_7H_{14} + 10\tfrac{1}{2}O_2 \rightarrow 7CO_2 + 7H_2O$$

Alcohols and ethers

Chemical Ideas 13.2

Alcohols

- Have a *hydroxyl* (–OH) group and have names ending in *-ol*.

Naming alcohols

- Count the number of carbons in the longest chain.
- Replace the *e* at the end of the parent alkane with *ol* – but retain the e for a *diol* or *triol*. For example, CH_2OHCH_2OH is ethane-1,2-diol.
- Locate the position of the OH group with as *low* a number as possible.

Formulae

Propan-1-ol has a full structural formula of

H H H
| | |
H — C — C — C — O — H, a shortened
| | |
H H H

structural formula of $CH_3CH_2CH_2OH$ and a skeletal formula of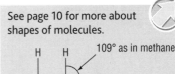

Combustion

- Alcohols react completely with oxygen to give carbon dioxide, CO_2, and water, H_2O. For example,

$$CH_3OH + 1\tfrac{1}{2}O_2 \rightarrow CO_2 + 2H_2O$$

Oxygenates

- Alcohols require less oxygen for complete combustion than the corresponding alkane, since the molecules *already have oxygen in them*. Compare these:

$$C_2H_5OH + 3O_2 \rightarrow 2CO_2 + 3H_2O$$
$$C_2H_6 + 3\tfrac{1}{2}O_2 \rightarrow 2CO_2 + 3H_2O$$

- Alcohols (and ethers) are called **oxygenates**. These burn more efficiently than alkanes, producing less carbon monoxide. They are commonly added to petrol to reduce pollution.

Ethers

Have an *alkoxy* group, –OR. For example,

H H H   H
| | | |
H — C — C — C — O — C — H
| | | |
H H H   H

This is methoxypropane and *not* propoxymethane. The prefix is the shorter carbon chain.

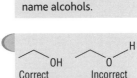

You might like to revise naming alkanes (page 24) before starting to name alcohols.

Correct | Incorrect

You will lose marks in an exam question if you use the incorrect version of a skeletal structure.

Number of carbons	Alcohol
1	methanol
2	ethanol
3	propanol
4	butanol
5	pentanol
6	hexanol

See page 10 for more about shapes of molecules.

109° as in methane
109° as in water

Alkoxy groups.

OCH_3	methoxy
OC_2H_5	ethoxy
OC_3H_7	propoxy

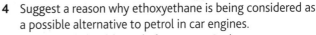

QUICK CHECK QUESTIONS

1 Draw the full structural formulae of methanol and ethanol.
2 Name this alcohol:

H H H H H
| | | | |
H — C — C — C — C — C — H
| | | | |
H H H OH H

3 Draw the full structural formulae for these alcohols and give their names.

(a)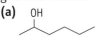

(b)

4 Suggest a reason why ethoxyethane is being considered as a possible alternative to petrol in car engines.
5 Draw the skeletal formula for propan-2-ol.
6 Write a balanced equation for the complete combustion of propan-1-ol, $CH_3CH_2CH_2OH$.

23

Alkanes and other hydrocarbons

Chemical Ideas 12.1, 12.2, 12.3 and Chemical Storylines DF3

Crude oil

Crude oil is a mixture of hundreds of different hydrocarbons. **Hydrocarbons** are compounds of carbon and hydrogen *only*. The hydrocarbons are separated into fractions by the process of *fractional distillation*. A *fraction* is a mixture of compounds with a specific boiling point range. These fractions are used as fuels.

Name of fraction	Boiling point range (°C)	Number of carbons in hydrocarbons	Use
gasoline	25 to 75	between 5 and 7	petrol
kerosene	190 to 250	between 10 and 16	jet fuel
gas oil	250 to 350	between 14 and 20	diesel

All the angles in alkanes are the same, 109°. The bonds are arranged tetrahedrally

Open-chain alkanes

- Have the general formula C_nH_{2n+2}.
- Name ends in -*ane*.
- Are **saturated** – all the bonds between carbon atoms are *single bonds*.
- Are **aliphatic** – they have no ring structures.

> Double check to ensure that you choose the longest chain.

Naming alkanes

- Choose the longest carbon chain and name it.
- Use prefixes in alphabetical order for any alkyl side chains.
- Use *di*, *tri*, *tetra* before the prefix if the side chains are identical.
- Show the position of any side chains by using numbers which are as low as possible.

Examples

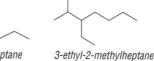

2,2-dimethylheptane   3-ethyl-2-methylheptane

Put a comma between numbers and a hyphen between a number and a letter.

Burning alkanes

- Alkanes react completely with oxygen to produce carbon dioxide and water. This is called complete combustion.

$$C_3H_8 + 5O_2 \rightarrow 3CO_2 + 4H_2O$$

Other reactions

- Alkanes also undergo radical substitution reactions. See page 58.

Learn these names.

CH_4	**meth**ane
C_2H_6	**eth**ane
C_3H_8	**prop**ane
C_4H_{10}	**but**ane
C_5H_{12}	**pent**ane
C_6H_{14}	**hex**ane
C_7H_{16}	**hept**ane
C_8H_{18}	**oct**ane
C_9H_{20}	**non**ane
$C_{10}H_{22}$	**dec**ane

Different types of formulae

Shortened structural formula	Full structural formula		Skeletal formula	
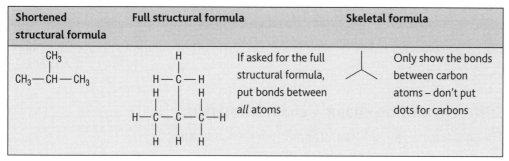		If asked for the full structural formula, put bonds between *all* atoms		Only show the bonds between carbon atoms – don't put dots for carbons

Don't forget to check that each carbon atom in a full structural formula has four bonds coming from it.

Skeletal formulae are most often used for very complicated structures, not simple structures such as ethane and propane.

Other hydrocarbons

Cycloalkanes

- Have the general formula C_nH_{2n}.
- Name ends in *-ane*.
- Are **saturated** – all the bonds between carbon atoms are *single bonds*.
- Are *not* **aromatic** – they don't have a benzene ring.

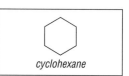
cyclohexane

Alkenes

- Have the general formula C_nH_{2n}.
- Name ends in *-ene*.
- Are **unsaturated** – there are one (or more) *double bonds* between carbon atoms in a molecule.
- Are **aliphatic** – they don't have ring structures.

propene

Alkenes undergo electrophilic addition reactions. See pages 66 and 67.

Arenes

- Name ends in *-ene*.
- Are **unsaturated**.
- Are **aromatic** – they have one or more *benzene rings* in their structure

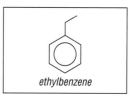
ethylbenzene

QUICK CHECK QUESTIONS

1 Name $CH_3–CH_2–CH_2–CH_2–CH_2–CH_2–CH_2–CH_3$.
2 Draw the skeletal formulae for 3-methylhexane and 2,2,4-trimethylheptane.
3 Write balanced equations for the complete combustion of methane and of ethane.
4 What do the molecular formulae of hexene, 2-methylpentene and cyclohexane have in common?
5 Draw the skeletal formula for the cycloalkane C_5H_{10}. Why is it aliphatic and not aromatic?

6 Name this cycloalkane:

7 What is the molecular formula of an alkane with 10 carbon atoms?
8 What is the molecular formula of a cycloalkane with 10 carbon atoms?

Structural isomerism

Chemical Ideas 3.3

Structural isomers have the same molecular formula but different structural formulae.

There are three main types of structural isomerism.

Different carbon chains – chain isomerism

The chain lengths are different. This is often seen in alkanes.

butane, C_4H_{10} *2-methylpropane*, C_4H_{10}

Different positions for the functional group – positional isomerism

The same functional group appears in different positions. This is often seen in alcohols.

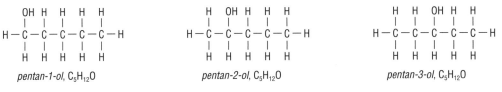

pentan-1-ol, $C_5H_{12}O$ *pentan-2-ol*, $C_5H_{12}O$ *pentan-3-ol*, $C_5H_{12}O$

Different functional groups – functional group isomerism

The molecular formula are the same but the functional groups are different. This can be seen in alcohols and ethers.

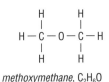

ethanol, C_2H_6O *methoxymethane*, C_2H_6O

Isomerisation at the oil refinery

When straight-chain alkanes are heated in the presence of a platinum catalyst, they become branched alkanes. The branched-chain alkanes have a higher *octane number* and less tendency to auto-ignite.

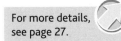
For more details, see page 27.

QUICK CHECK QUESTIONS

1 (a) Give the skeletal formulae of the isomers 2-methylpropan-2-ol and 2-methylpropan-1-ol.
   (b) Why are they said to be isomers?

2 Give the full structural formulae for two isomers with the molecular formula C_2H_6O. Name the functional group in each of them.

3 Which two of these structures represent the same compound?

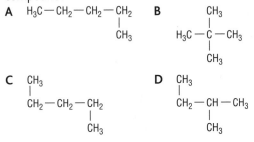

Auto-ignition and octane numbers

Chemical Storylines DF3 and DF4

Definitions

Auto-ignition is the explosion of a fuel without a spark. The **octane number** is a measure of the tendency of petrol to auto-ignite. Auto-ignition in a petrol engine causes:

- a knocking or 'pinking' sound
- reduced engine performance
- engine damage.

Octane numbers

If a fuel has an octane number of 80, it has the same tendency to auto-ignite as a mixture of 80% 2,2,4-trimethylpentane and 20% heptane.

100% 2,2,4-trimethylpentane		100% heptane
octane number 100	$\longrightarrow$	octane number 0
low tendency to auto-ignite		high tendency to auto-ignite

> Remember that a *high* octane number means a *low* tendency to auto-ignite.

Comparing octane numbers

- Short-chain compounds have a higher octane number than long-chain compounds.
- Branched-chain compounds have a higher octane number than the corresponding unbranched-chain compounds.
- Cycloalkanes have higher octane numbers than the corresponding straight-chain alkanes.
- Arenes have higher octane numbers than cycloalkanes.
- Oxygenates have higher octane numbers than the corresponding alkanes and are added to petrol to increase the octane number.

The petrochemical industry uses a number of different chemical reactions, cracking, reforming and isomerisation, to increase the octane numbers of the components in petrol.

> Oxygenates, such as MTBE, have oxygen atoms in the molecule.
>
>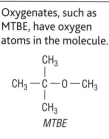
> *MTBE*
>
> Other examples of oxygenates are the alcohols, methanol and ethanol.

> For more details see pages 26 and 28.

QUICK CHECK QUESTIONS

1 Put the following compounds in order of increasing octane number.

benzene

hexane

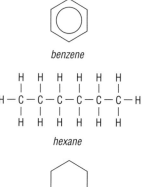

cyclohexane

2 Which of these two compounds has the lowest octane number? How do you know this?

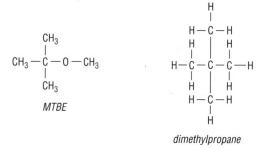

MTBE

dimethylpropane

3 Unblended gasoline has an octane number of 70. The octane rating of ethanol is 111 and can be blended with petrol to increase octane numbers. How does the higher octane number affect the performance of the fuel?

Catalysts and altering octane numbers

Chemical Ideas 10.5, Chemical Storylines DF4 and DF7

- A **catalyst** speeds up a chemical reaction but is not used up in the reaction, this process is called **catalysis**.
- **Heterogeneous catalysts** are in a different physical state to the reactants.

In a catalytic converter in a car, platinum or rhodium act as heterogeneous catalysts for the following reaction:

$$2NO(g) + 2CO(g) \rightarrow N_2(g) + 2CO_2(g)$$

Catalysts in catalytic converters only work when the temperature is high. They don't work at the start of a journey when a car engine is cold.

If state symbols are required in the answer to a question, you will be told in the question.

A catalyst poison adsorbs onto the catalyst surface and stops it working.

How heterogeneous catalysts work
1 Reactants are *adsorbed* onto the surface of the catalyst (or bond to the surface).
2 Bonds between atoms inside the reactants *weaken* and are *broken*.
3 New bonds and compounds are *formed*.
4 The products *diffuse* away.

Catalysts and petrol components

Catalysts are used in three processes at the oil refinery to produce petrol components with a *low tendency* to auto-ignite. The products of these three processes have a *higher octane number* than the reactants in the processes.

Cracking

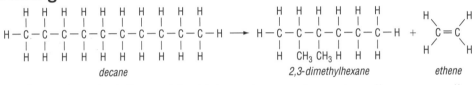

decane → 2,3-dimethylhexane + ethene

- A *shorter*, *branched* alkane with a higher octane number and an *alkene* are usually formed.
- A zeolite is used as the catalyst.

Reforming

- Alkanes are converted to cycloalkanes and hydrogen.
- Cycloalkanes are then converted to arenes and hydrogen.
- In terms of octane numbers, *arenes > cycloalkanes > alkanes*.
- Platinum is used as the catalyst.

Isomerisation

The product is shorter and more branched than the reactant.

$$CH_3 - CH_2 - CH_2 - CH_2 - CH_3 \quad \xrightarrow[\text{Zeolite sieve}]{\text{Pt catalyst}} \quad CH_3 - \overset{\displaystyle CH_3}{\underset{\displaystyle |}{CH}} - CH_2 - CH_3$$

pentane, octane number 62   2-methylbutane, octane number 93

- The longest carbon chain becomes *shorter* so the octane number increases.
- There is more *branching* so the octane number increases.
- The zeolite isn't a catalyst here; it is a sieve to separate the branched and unbranched isomers.
- Platinum is the catalyst for these reactions.

QUICK CHECK QUESTIONS

1 Write a balanced chemical equation (using molecular formulae) for the cracking of octane to give an alkene with three carbon atoms and an alkane.

2 Is this reaction cracking, reforming or isomerisation?

$$C_7H_{16} \rightarrow C_7H_{14} + H_2$$

3 When carbon monoxide and nitrogen monoxide in a car exhaust pass through a catalytic converter, carbon dioxide and nitrogen are produced. Write a balanced equation for this reaction, including state symbols.

Pollution from cars

Chemical Storylines DF6–DF10

Production and effects of pollutants from petrol

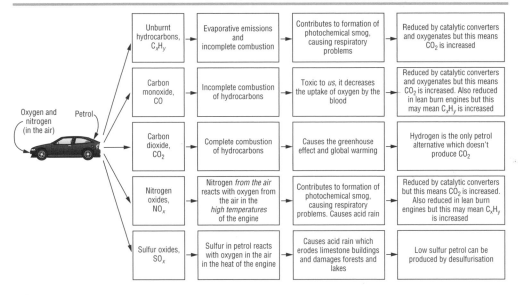

It is important to say that the CO is harmful to humans.

Alcohols and ethers are oxygenates. Since they have oxygen atoms in them, their combustion is more complete than the combustion of the corresponding alkane. Less CO and C_xH_y are produced.

It is important to mention that in the production of oxides of nitrogen in a car engine, the nitrogen and oxygen are both *from the air* and that *high temperatures* are required.

$$N_2 + O_2 \rightarrow 2NO$$

Alternative fuels for a car engine

People need to make choices, both now and in the future, regarding the use of fuels. We need to weigh up the risks, benefits and sustainability for each fuel.

Energy density (kJ kg^{-1}) = enthalpy of combustion (kJ mol^{-1}) × number of moles of fuel in 1 kg (mol kg^{-1}).

Alternative	Is it sustainable?	Benefits	Risks
Diesel	No; crude oil is running out	Less CO produced than from a petrol engine; already sold at petrol stations	Produce more NO_x and particulates than a petrol engine; particulates can irritate lungs
LPG or autogas	No; crude oil is running out	Less CO, CO_2, C_xH_y and NO_x than from a petrol engine; petrol engines easily converted	Needs to be stored under pressure so that it is a liquid
Ethanol	Possibly not; large amounts of energy needed for cultivating sugar cane for fermentation	Less CO, SO_2 and NO_x than from a car engine; ethanol has a high octane number; sugar cane absorbs CO_2 in growth	Highly flammable
Biodiesel	Can be made from waste plant and animal oils and fats so renewable but fossil fuels may be used as an energy source in production	Living things have absorbed CO_2; it is biodegradable; less CO, C_xH_y, SO_2 and particulates than from a diesel engine	NO_x emissions higher than a diesel engine
Hydrogen	Only if the electricity needed for electrolysis of water is from a renewable source such as solar cells	Water is the only product of combustion	Highly flammable; high pressure fuel tank needed to store it as a liquid

QUICK CHECK QUESTIONS

1 Explain how carbon monoxide is produced in cars.
2 Describe one polluting effect of nitrogen monoxide.
3 What problems may arise in the storage of hydrogen in a fuel tank?
4 Biodiesel can be manufactured from soya beans. Give one possible advantage and one disadvantage of such a fuel.

5 A 'lean burn' engine uses a higher ratio of air to petrol vapour than other engines. Why are less unburnt hydrocarbons produced?
6 Biodiesel molecules contain oxygen atoms. What name is given to molecules such as these which are added to fuels to reduce certain types of pollution?

Unit F331
Practice exam-style questions

Note: Each of these practice questions covers a single teaching module (e.g. **Elements of Life**). In your actual exams, each question may cover more than one teaching module.

Elements of Life (EL)

1 Nuclear fusion reactions are common in the centres of stars, where temperatures can reach hundreds of millions of degrees.

(a) (i) What is meant by the term *nuclear fusion*? [2]

(ii) Why are such high temperatures needed for this type of reaction? [1]

(b) Our own Sun is a lightweight star and is made up mainly of hydrogen.

(i) Complete the following nuclear equation to show the formation of a new element:

$$_1^1H + {}_1^2H \rightarrow \; _{-}^{-} \text{-----}$$ [2]

(ii) What term is used to describe two forms of the same element, such as ^1H and ^2H? [1]

(iii) Describe how the two forms of hydrogen are different. [1]

(c) Scientists are working towards using nuclear fusion as an alternative energy source. One process uses a form of hydrogen called tritium, ^3H. Tritium is radioactive and has a half-life of 12.25 years.

(i) Tritium decays by β emission. Describe **two** properties of β radiation. [2]

(ii) Explain why the atoms of some elements are radioactive. [1]

(iii) 100 g of tritium decays. How much tritium would be left after 49 years? [2]

(d) Scientists have been able to determine which elements are present in stars, and elsewhere in the Universe, by using spectroscopic methods.

Explain how the atomic absorption spectrum of a star can be used to identify the elements that it contains.

QWC: In your answer, you should use appropriate technical terms, spelled correctly. [4]

(e) The space between stars is called the interstellar medium. This is largely empty space but spectroscopic evidence shows the presence of some molecules. One of the molecules that has been identified is ammonia, NH_3.

(i) Write out the electron structure of a nitrogen atom. [1]

(ii) Draw an electron dot–cross diagram for an ammonia molecule. [2]

(iii) Draw a diagram to show the shape of an ammonia molecule. Include a value for the H–N–H bond angle. Explain your answer. [5]

(f) Another nitrogen-containing molecule has recently been found in the interstellar medium. Analysis has shown that this molecule has the following percentage composition by mass: C 38.7%, H 16.1%, N 45.2%

(i) Calculate the empirical formula of this compound. [3]

(ii) Explain how a mass spectrum could be used to determine the molecular formula of this nitrogen-containing molecule. [2]

[Total: 29]

Developing Fuels (DF)

2 Oil companies spend large sums of money on research into alternative fuels to petrol. 'Biobutanol', C_4H_9OH, can be produced by fermentation of sugar beet.

(a) (i) Draw the *full structural* formula of butan-1-ol. [2]

Show all bonds in the molecule.

    (ii) To which *homologous series* does butan-1-ol belong? [1]

(b) Biobutanol has several advantages compared to other fuels. One advantage is that biobutanol is an *oxygenate*.

    (i) Explain one advantage of using *oxygenates* such as biobutanol in fuels. [1]

    (ii) Suggest **one other** advantage of biobutanol compared to petrol. [1]

A full definition is required for 3 marks

(c) (i) Explain what is meant by *standard enthalpy change of combustion*. [3]

    (ii) Complete the enthalpy cycle below, which includes the standard enthalpy of combustion of butan-1-ol and its standard enthalpy of formation. [2]

Start by writing the equation for the enthalpy change requested.

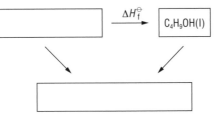

    (iii) Use the enthalpy cycle from **(c)(ii)** and the data on the right to calculate a value for the enthalpy change of formation of butan-1-ol.

Your answer should have a sign and units.

Substance	$\Delta H_c^{\ominus}$ (kJ mol⁻¹)
C(s)	−394
$H_2(g)$	−286
$C_4H_9OH(l)$	−2673

[3]

(d) Another alternative fuel to petrol is liquid petroleum gas (LPG). This is a mixture of propane, C_3H_8, and butane, C_4H_{10}. LPG is kept as a liquid, under pressure, in a reinforced tank.

    (i) Draw the *skeletal* formulae for propane and butane. [2]

C–C bonds are shown as lines and C–H bonds are not shown.

    (ii) Write a balanced equation for the complete combustion of propane. Include state symbols. [2]

    (iii) Use the equation in **(d)(ii)**, and the data in the table, to calculate a value for the enthalpy change of combustion, ΔH_c, of propane.

Average bond enthalpies (kJ mol⁻¹)	
C–C	+348
C–H	+412
O–H	+464
C=O	+805
O=O	+498

[4]

    (iv) The Data Book value for the standard enthalpy change of combustion ($\Delta H_c^{\ominus}$) of propane is −2220 kJ mol⁻¹. Suggest **one** reason why your calculated value is different from the Data Book value. [1]

(e) An experiment was carried out to find the relative molecular mass of a sample of LPG. 1 dm³ of the LPG had a mass of 2.25 g at room temperature and pressure.

    (i) Calculate a value for the relative molecular mass of this sample of LPG. [1]
        Assume that 1 mole of gas occupies 24 dm³ at room temperature and pressure.

    (ii) Suggest what the value for the relative molecular mass tells you about the gas mixture in this sample of LPG. [2]

First step is finding the M_r of butane.

[Total: 25]

Developing Fuels (DF)

3 In the United States of America, the fuel additive known as MTBE has been widely used in petrol blends to increase the octane rating of fuels.

(a) (i) What is meant by the term *octane number*. [2]

(ii) Suggest **one** important reason for ensuring that a fuel has the correct octane rating for the engine in which it is being used. [1]

(iii) MTBE has the formula

$$H_3C-\overset{\overset{\displaystyle CH_3}{|}}{\underset{\underset{\displaystyle CH_3}{|}}{C}}\overset{x}{\diagup}O\diagdown_{y}CH_3$$

What are the values of the two bond angles *x* and *y*? [2]

> Repulsion of electron pairs determines bond angles.

(b) In the UK, further refining of alkanes is used to increase the octane rating of fuels. These refinery processes include *isomerisation*, *reforming* and *cracking*.

(i) Describe how isomerisation produces alkanes that have higher octane numbers. [2]

(ii) Draw the structural formula of one possible product of the isomerisation of hexane, C_6H_{14}. Give the name of the hydrocarbon you have drawn. [3]

> Make sure the formula for your product is correct and that the name matches the structure drawn.

(iii) Describe how zeolites are used to separate the products from the unchanged reactants in the isomerisation process. [1]

(c) The reforming process converts straight-chain alkanes into ring compounds. Hot alkane vapours are passed over a catalyst. The catalyst used in the process is platinum metal, which is finely dispersed on a solid support of aluminium oxide. Name the *type* of catalyst used in the reforming process. [1]

> There are two *types* of catalyst – homogeneous and heterogeneous.
> Homo means 'same'; hetero means 'different'.

(d) (i) Complete the table below, by placing a tick (✓) in the appropriate boxes, to show which refining process produces each of the hydrocarbon fuel components. [2]

Component	Refining process		
	Cracking	Reforming	Straight-run
heptane			
hex-1-ene			
benzene (an arene)			
cyclohexane			

(ii) Which of the hydrocarbons in **(d)(i)** can be described as *unsaturated*? [1]

> Entropy is affected by the state of matter of a substance.
> Entropy is a measure of the number ways of arranging particles.

(e) Petrol blends contain a mixture of gases such as butane, and liquids such as cyclohexane. Which has the higher entropy, butane or cyclohexane? Explain your answer. [2]

(f) All petrol-engined vehicles produce pollutants in their exhaust emissions, no matter which fuel blend is used. Two of these pollutants are nitrogen monoxide, NO, and unburnt hydrocarbons.

(i) Give the names of **two other** pollutants from petrol-engined vehicles. [2]

(ii) Give **one** reason why it is desirable to reduce hydrocarbon emissions. [1]

(iii) Explain how nitrogen monoxide forms in a car engine. [2]

[Total: 22]

Elements from the Sea (ES)

This module looks at chemistry of the halogens in the oceans and industry. This introduces redox reactions, intermolecular bonding, halogenoalkanes and greener industry.

In exam questions on this module, you may also be expected to answer questions using ideas already met in previous teaching modules for Unit F331. These include using formulae (page 2), drawing molecular shapes in 3D (page 10), predicting molecular shapes (page 10), doing calculations involving masses and volumes of gases (pages 17 and 18), using dot–cross diagrams (pages 8 and 9) and recalling the physical properties of giant lattices and simple molecules (pages 8 and 9).

As you work through this module, you will cover the following concepts. 'CI' refers to sections in your *Chemical Ideas* textbook. 'Storylines' refers to your *Chemical Storylines* textbook.

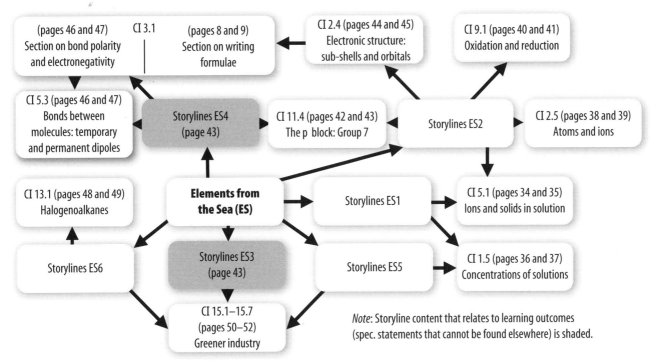

Note: Storyline content that relates to learning outcomes (spec. statements that cannot be found elsewhere) is shaded.

From *Chemical Storylines*, you also need to be aware of the following:

Balancing benefits and risks of halogens

Halogen-containing compounds have brought huge benefits to society, but there are some significant problems associated with the elements and their compounds.

Halogen	Benefits	Risks
Fluorine	Used to make plastics such as PTFE. Added to toothpaste to strengthen tooth enamel. Used to make HCFCs.	The element itself is highly reactive and handling must be kept to a minimum.
Chlorine	An important intermediate in the manufacture of hydrochloric acid and chlorinated solvents. Used in the plastics industry (for PVC and polyurethanes). Used in water treatment and to make pesticides, medicines and bleach.	Pesticides can accumulate in the environment (e.g. DDT). CFCs destroy stratospheric ozone.
Bromine	Used in the manufacture of flame retardants, agricultural fumigants and in photography.	Organic bromine compounds can destroy ozone in the stratosphere.
Iodine	Used in antiseptics, germicides and dyes. Iodine-131 is used to diagnose thyroid disease.	

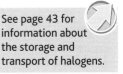

See page 43 for information about the storage and transport of halogens.

Ions in solids and solutions

Chemical Ideas 5.1

Structure of an ionic lattice: sodium chloride

You may be asked to draw the structure of an ionic compound. Remember to label the ions and their charges.

Not all ionic solids have simple cubic lattices. The type of lattice depends on the number and size of the anions and cations.

Sodium chloride consists of sodium ions (Na^+), each of which is surrounded by six chloride ions (Cl^-). In turn, each Cl^- ion is surrounded by six Na^+ ions. The ions are held together by attractions between these oppositely charged ions. This allows a **giant ionic lattice** to be built up. Similar electrostatic bonds hold all ionic lattices together. The lattice structure of sodium chloride is said to be **simple cubic**.

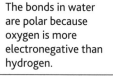

Chloride ions

Sodium ions

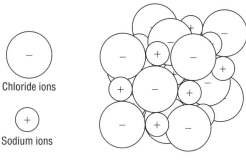

Sodium chloride lattice.

Sometimes, ionic crystals contain some water molecules. These water molecules sit within the ionic lattice and are known as **water of crystallisation**. The crystals are said to be **hydrated**. You may have seen blue copper sulfate crystals – these are in fact the pentahydrate crystal, $CuSO_4 \cdot 5H_2O$. Anhydrous copper sulfate ($CuSO_4$) is white.

Ionic substances in solution

The bonds in water are polar because oxygen is more electronegative than hydrogen.

Many ionic compounds dissolve in water without difficulty. As they dissolve, the ions become surrounded by water molecules and they spread throughout the solution. The hydrated ions are *randomly* arranged and behave *independently*.

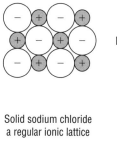

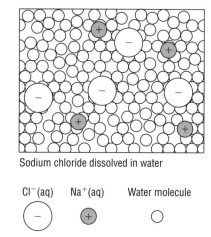

Solid sodium chloride
a regular ionic lattice

Sodium chloride dissolved in water

Cl^- $\qquad$ Na^+ $\qquad\qquad$ $Cl^-(aq)$ $\qquad$ $Na^+(aq)$ $\qquad$ Water molecule

Sodium chloride dissolving.

A hydrated ion is represented by the formula of that ion, followed by (aq), e.g. $Cl^-(aq)$.

If you are asked to draw a labelled diagram of a hydrated ion always label the charge on the ion, show the polarity of the water molecules and surround the ion with a minimum of five water molecules.

Water surrounds ions in the way that it does because it is a polar molecule with a bent shape. The positive hydrogen atoms in water are attracted to negative ions. Water molecules surround positive ions with the oxygen atoms pointing inwards. Each ion is surrounded by its own sphere of water molecules – this process is known as **hydration**.

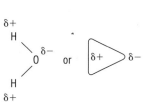

Polarity in a water molecule.

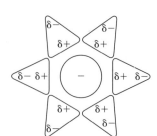

Hydrated negative ion.

Hydrated positive ion.

Ionic equations

Ions in solution behave independently – and this includes when they are involved in chemical reactions. If two solutions react to form a solid, a **precipitation** reaction is said to have occurred. For example:

$$Ag^+(aq) + NO_3^-(aq) + Na^+(aq) + Br^-(aq) \rightarrow AgBr(s) + Na^+(aq) + NO_3^-(aq)$$

$Na^+(aq)$ and $NO_3^-(aq)$ are **spectator ions** (i.e. they are not involved in the reaction) and are not included in the **ionic equation**, which is written:

$$Ag^+(aq) + Br^-(aq) \rightarrow AgBr(s)$$

> Always put in the state symbols for an ionic equation.

WORKED EXAMPLE

Write an ionic equation for the reaction of barium chloride solution with sodium sulfate solution. A precipitate of barium sulfate is formed.

STEP 1 Write down symbols for the ions in each solution and the ions in the product.

$$Ba^{2+}(aq) + 2Cl^-(aq) + 2Na^+(aq) + SO_4^{2-}(aq) \rightarrow BaSO_4(s) + 2Na^+(aq) + 2Cl^-(aq)$$

STEP 2 Cross out those ions which are the same on the left and right of the arrow. This leaves the ionic equation:

$$Ba^{2+}(aq) + SO_4^{2-}(aq) \rightarrow BaSO_4(s)$$

The following rules will help you to predict if ionic precipitation reactions will take place when two solutions are mixed.

- all nitrates are soluble in water
- all chlorides are soluble in water except AgCl and $PbCl_2$
- all sulfates are soluble in water except $BaSO_4$, $PbSO_4$ and $SrSO_4$
- all sodium, potassium and ammonium salts are soluble in water
- all carbonates are insoluble in water except $(NH_4)_2CO_3$ and those of the Group 1 elements.

Neutralisation reactions can also be summarised using ionic equations. When an acid and alkali react, a covalent compound is formed – in this case water. When hydrochloric acid reacts with sodium hydroxide the ions involved, and the products, are:

$$H^+(aq) + Cl^-(aq) + Na^+(aq) + OH^-(aq) \rightarrow H_2O(l) + Na^+(aq) + Cl^-(aq)$$

When the spectator ions have been removed, the ionic equation is:

$$H^+(aq) + OH^-(aq) \rightarrow H_2O(l)$$

> *All* acid/alkali neutralisation reactions have this ionic equation, whatever acid and alkali are involved.

QUICK CHECK QUESTIONS

1 Write down the ions present in solutions of the following compounds:
   (a) Na_2SO_4 (b) $CaCl_2$ (c) $Fe(NO_3)_3$
2 Write the formula of the following ionic compounds:
   (a) magnesium fluoride
   (b) aluminium oxide
   (c) sodium carbonate.
3 Write ionic equations, with state symbols, for the following precipitation reactions:
   (a) lead nitrate + sodium sulfate →
                    lead sulfate + sodium nitrate
   (b) silver nitrate + magnesium chloride →
                    silver chloride + magnesium nitrate.

4 (a) Write an equation showing all the ions for the neutralisation reaction between sulfuric acid and potassium hydroxide.
   (b) Explain how to obtain the ionic equation for this reaction.
5 What structure is the lattice of sodium chloride?
6 Draw a diagram of a potassium ion hydrated by five water molecules, clearly showing the direction of the dipole in water.
7 Give the name of the precipitate that would be formed when the following solutions are mixed:
   (a) ammonium carbonate and barium chloride
   (b) lead nitrate and sodium chloride.

Concentrations of solutions

Chemical Ideas 1.5

Calculations of concentrations

Concentrations can be measured in grams per cubic decimetre ($g\,dm^{-3}$).

Example

$1\,dm^3 = 1000\,cm^3$	

A solution containing 40 g of sodium hydroxide in $2\,dm^3$ of solution has a concentration of

$$\frac{40}{2} = 20\,g\,dm^{-3}.$$

However, chemists prefer to measure quantities in moles – the number of particles present. A more common unit for concentration is *moles per cubic decimetre* ($mol\,dm^{-3}$).

1 mole of a substance, dissolved and then made up to a volume of $1\,dm^3$ of solution, has a concentration of $1\,mol\,dm^{-3}$.

> 1 mole of a substance is equal to its relative formula mass (M_r) in grams.

> Always remember to convert volumes to dm^3.
>
> Divide cm^3 by 1000 to get the volume in dm^3.

Example

A memory aid can be used to help with concentration calculations.

- concentration (c) in $mol\,dm^{-3}$
- amount of substance (n) in moles
- volume of solution (V) in dm^3

$$c = \frac{n}{V} \qquad n = c \times V \qquad V = \frac{n}{c}$$

Any of the three quantities (concentration, amount or volume) can be calculated by using the correct expression and the other two known values.

Examples

> $100\,cm^3$ is $0.1\,dm^3$
> $20\,cm^3$ is $0.02\,dm^3$

Concentration	Amount	Volume
$c = \dfrac{n}{V}$	$n = c \times V$	$V = \dfrac{n}{c}$
What is the concentration of a solution of 0.5 mole of NaOH in $100\,cm^3$?	How many moles in $20\,cm^3$ of $0.1\,mol\,dm^{-3}$ NaOH?	What volume of $0.2\,mol\,dm^{-3}$ NaOH contains 0.1 mol?
$c = \dfrac{0.5}{0.1}$	$n = 0.1 \times 0.02$ $= 0.002\,mol$	$V = \dfrac{0.1}{0.2}$
$= 5\,mol\,dm^{-3}$		$= 0.5\,dm^3$
unit: $mol\,dm^{-3}$	unit: mol	unit: dm^3

> Do not forget the units.

Using concentrations in calculations

A titration is a method of quantitatively finding the concentration of a solution by reacting a known volume of it with another solution of known concentration. The *end-point* of the reaction is often detected by the use of an indicator that changes colour.

- A fixed volume of solution of unknown concentration is placed in a conical flask using a pipette.

- A few drops of a suitable indicator are added and the mixture placed on a white tile (to see the end-point clearly).
- The solution of known concentration is added slowly from a burette, with constant swirling.
- As you approach the end-point, add the solution from the burette *dropwise*.
- After a rough titration, accurate titrations follow until **concordant** results are obtained.

See page vi for more about experimental techniques.

'Concordant' means consistent. Aim to get your accurate titres to within $0.10\,cm^3$.

WORKED EXAMPLE

In a titration, $25.0\,cm^3$ of potassium hydroxide solution was pipetted into a conical flask. A $0.020\,mol\,dm^{-3}$ solution of sulfuric acid was added from a burette. An indicator in the solution changed colour when $27.9\,cm^3$ of sulfuric acid had been added. What is the concentration of the potassium hydroxide?

STEP 1 Write down the equation:

$$H_2SO_4 + 2KOH \rightarrow K_2SO_4 + 2H_2O$$

STEP 2 Find the ratio of the reactants:

1 mole of sulfuric acid reacts with 2 moles of potassium hydroxide.

STEP 3 Using the equation $n = c \times V$, you can calculate the number of moles of H_2SO_4:

$$n(H_2SO_4) = 0.020 \times \frac{27.9}{1000} = 5.58 \times 10^{-4}\,mol$$

STEP 4 Use the ratio from step 2:

This is equivalent to $2 \times 5.58 \times 10^{-4} = 1.116 \times 10^{-3}$ moles of KOH.

STEP 5 Calculate the concentration:

This number of moles is in a volume of $25.0\,cm^3$. You can calculate the concentration of the KOH solution from the equation

$$c = \frac{n}{V} = \frac{1.116 \times 10^{-3}}{0.025} = 0.045\,mol\,dm^{-3}$$

Always quote your final result to the same number of significant figures as the data with the lowest number of significant figures quoted in the question.

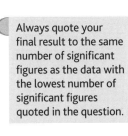

Volume of H_2SO_4
$= 27.9\,cm^3 = \frac{27.9}{1000}\,dm^3$

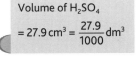

1 mole of H_2SO_4 reacts with 2 moles of KOH.

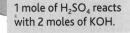

Volume of KOH
$= 25.0\,cm^3 = \frac{25.0}{1000}\,dm^3$
$= 0.0250\,dm^3$

QUICK CHECK QUESTIONS

1 Convert the following volumes into dm^3:
   (a) $50\,cm^3$ (b) $2500\,cm^3$
2 How many moles of solute are contained in each of the following solutions:
   (a) $5\,dm^3$ of $0.1\,mol\,dm^{-3}$ NaOH
   (b) $200\,cm^3$ of $2.0\,mol\,dm^{-3}$ H_2SO_4?
3 What is the concentration of the solutions in question **2** in $g\,dm^{-3}$?
4 What is the concentration (in $mol\,dm^{-3}$) of the following solutions:
   (a) $40\,g$ of NaOH in $500\,cm^3$ of solution
   (b) $10.6\,g$ of Na_2CO_3 in $2000\,cm^3$ of solution?
5 What mass of solute is needed to make up the following solutions:
   (a) $100\,cm^3$ of a $0.5\,mol\,dm^{-3}$ solution of $MgCl_2$
   (b) $2\,dm^3$ of a $0.02\,mol\,dm^{-3}$ solution of $KMnO_4$?
6 What pieces of apparatus would you need to carry out a titration between solutions of sodium hydroxide and hydrochloric acid?

7 A volume of $20\,cm^3$ sulfuric acid was used in a titration, delivered from a burette. The concentration of the solution was $0.100\,mol\,dm^{-3}$.
   (a) What was the volume of acid used, in dm^3?
   (b) What was the number of moles of acid used?
   (c) How many moles of NaOH would this acid neutralise?

   $$H_2SO_4(aq) + 2NaOH(aq) \rightarrow Na_2SO_4(aq) + 2H_2O(l)$$

   (d) If this number of moles of NaOH were in a $25.0\,cm^3$ portion drawn by a pipette and placed in a conical flask, how many moles would there have been in $1000\,cm^3$ of the solution?
   (e) What is the concentration of the sodium hydroxide solution, in $mol\,dm^{-3}$?
8 $18.6\,cm^3$ of $0.10\,mol\,dm^{-3}$ nitric acid was required to neutralise $25.0\,cm^3$ of a solution of calcium hydroxide. Calculate the concentration of the calcium hydroxide solution.

Atoms and ions

Chemical Ideas 2.5

First ionisation enthalpy

The first ionisation enthalpy (or first ionisation energy) is the energy needed to remove one electron from each of *one mole* of *isolated gaseous* atoms of an element. *One mole of gaseous ions* with one positive charge is formed. The units are $kJ\,mol^{-1}$.

$$X(g) \rightarrow X^+(g) + e^-$$

Remember to put (g) in ionisation enthalpy equations.

The value of the 1st ionisation enthalpy is always positive – energy must be put in to remove the electron because it is attracted to the nucleus.

The illustration below shows how the first ionisation enthalpy varies with atomic number for elements 1–56.

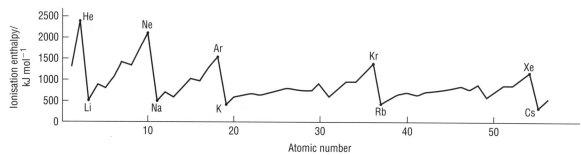

You should notice the following patterns:

These patterns are examples of **periodicity**. For other examples of periodicity see page 11.

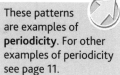

- The elements at the peaks are all in Group 0. It is difficult to remove an electron from these atoms with full outer shells and the elements are all very unreactive.
- The elements at the troughs are all in Group 1 (alkali metals). These atoms have only one outer shell electron and are relatively easy to ionise. They are all very reactive elements.

How does first ionisation enthalpy vary across a period?

The bar chart on the right shows the first ionisation enthalpy for elements in Period 2 (Li to Ne). The general trend is one of *increasing* first ionisation enthalpy as atomic number increases. The nuclear charge increases going across the period from left to right and the electrons are being added into the same shell. This means there is a greater attraction between the nucleus and the electron. Therefore, more energy is needed to remove the electron.

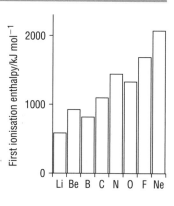

How does first ionisation enthalpy vary down a group?

Group 7 element	1st ionisation enthalpy (kJ mol⁻¹)
F	+1687
Cl	+1257
Br	+1146
I	+1010

First ionisation enthalpies always *decrease* as you go down a group. This is because the attraction between the nucleus and the outermost electrons decreases. There are more filled shells of electrons between the nucleus and the outermost electron and so:

- the outermost electron is further from the nucleus
- the filled electron shells **shield** the positively charged nucleus from the outermost electrons.

These two factors make it easier for an outermost electron to be removed.

Successive ionisation enthalpies

As well as first ionisation enthalpies, there are second, third, fourth etc. ionisation enthalpies which are the energies required to remove further electrons:

- first ionisation enthalpy: $X(g) \rightarrow X^+(g) + e^-$
- second ionisation enthalpy: $X^+(g) \rightarrow X^{2+}(g) + e^-$
- third ionisation enthalpy: $X^{2+}(g) \rightarrow X^{3+}(g) + e^-$
- fourth ionisation enthalpy: $X^{3+}(g) \rightarrow X^{4+}(g) + e^-$

> Take care NOT to write the following equation for the second ionisation enthalpy!
>
> $X(g) \rightarrow X^{2+}(g) + 2e^-$
>
> This an equation representing the first *and* second ionisations together.

The second and subsequent ionisations involve the removal of an electron from a positive ion.

Study the bar charts (right) which show successive ionisation enthalpies for the elements aluminium (Group 3) and phosphorus (Group 5). There are two important patterns to notice:

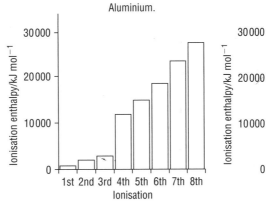

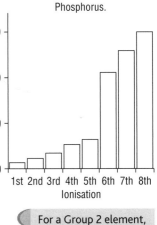

- Ionisation enthalpies increase as successive electrons are removed. After an electron is removed, the remaining electrons are attracted more strongly to the nucleus.
- There is a sharp jump in ionisation enthalpy when an electron is removed from a full electron shell. For example, an aluminium atom has electron arrangement 2.8.3. The fourth ionisation enthalpy for aluminium is much higher than the first, second or third because the fourth electron has to be removed from a full electron shell.

> For a Group 2 element, the sharp jump would occur between the 2nd and 3rd ionisation enthalpies.

QUICK CHECK QUESTIONS

1. Define the term *first ionisation enthalpy* in words.
2. Write equations to represent:
   (a) the first ionisation enthalpy of sodium
   (b) the second ionisation enthalpy of carbon.
3. Explain why the first ionisation enthalpy of phosphorus (atomic number 15) is higher than the first ionisation enthalpy of silicon (atomic number 14).
4. The first ionisation enthalpies for the first three elements in Group 1 are:

   Li $+519\,kJ\,mol^{-1}$; Na $+494\,kJ\,mol^{-1}$; Li $+418\,kJ\,mol^{-1}$

   (a) Why are all the values positive?
   (b) Explain the trend.
5. The graph below shows the first ionisation enthalpies for 16 elements in their order in the Periodic Table. The element at which the graph starts is not specified.

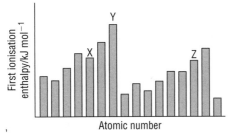

In which group of the periodic table is:
(a) element X
(b) element Y
(c) element Z?

Only answer the following two questions after you have studied *Electronic structure: sub-shells and orbitals* (pages 44 and 45).

6. Use electron arrangements to suggest why the first ionisation enthalpy of aluminium (atomic number 13) is slightly *lower* than the first ionisation enthalpy of magnesium (atomic number 12).
7. (a) Sketch a bar chart to show the first eight successive ionisation enthalpies for the element silicon, in Group 4 of the Periodic Table.

1st IE = $+768\,kJ\,mol^{-1}$,	2nd IE = $+1580\,kJ\,mol^{-1}$,
3rd IE = $+3230\,kJ\,mol^{-1}$,	4th IE = $+4360\,kJ\,mol^{-1}$,
5th IE = $+16\,000\,kJ\,mol^{-1}$,	6th IE = $+20\,000\,kJ\,mol^{-1}$,
7th IE = $+23\,600\,kJ\,mol^{-1}$,	8th IE = $+29\,100\,kJ\,mol^{-1}$

   (b) Why is there a large increase between the 4th and 5th ionisation enthalpies?

Oxidation and reduction

Chemical Ideas 9.1

When an oxidation reaction and a reduction reaction occur simultaneously, this is called a **redox reaction**.

Oxidation and reduction can be defined in two different ways:

- **o**xidation **is** the **l**oss of electrons.
- **r**eduction **is** the **g**ain of electrons.

or

- an element is oxidised when its oxidation state is increased (becomes more positive).
- an element is reduced when its oxidation state is decreased (becomes more negative).

> Remember **OIL RIG**: Oxidation Is Loss; Reduction Is Gain.

Assigning oxidation states

You should learn the following rules:

- The atoms in elements are always in an oxidation state of zero.
- In compounds or ions, oxidation states (or numbers) are assigned to each atom or ion. Since compounds have no overall charge, the oxidation states of all the constituents must add up to zero. In ions, the sum of the oxidation states is equal to the charge on the ion.
- Some atoms rarely change their oxidation states in reactions. These can be used to help to assign oxidation states to other species. Examples are F is -1, O is -2 (except in O_2^- and OF_2), H is $+1$, Cl is usually -1 (except when combined with O or F).

> You must always write down the sign of an oxidation state ($+$ or $-$) or you will lose marks. The sign is written *in front of* the oxidation number.

WORKED EXAMPLE

Assign the oxidation states for each element in the following compounds.

(a) CO_2

STEP 1 O is -2; there are two Os so the total contribution of O to the oxidation state is $2 \times (-2) = -4$.

STEP 2 To make the total of the oxidation states add up to zero, C must be $+4$.

(b) CH_4

STEP 1 H is $+1$; there are four so the total contribution of H to the oxidation state is $4 \times (+1) = +4$.

STEP 2 To make the total of the oxidation states add up to zero, C is -4.

(c) VO^{2+}

STEP 1 O is -2.

STEP 2 The charge on the ion is $+2$, so V must have an oxidation state of $+4$ because $+4 + (-2) = +2$.

Using oxidation states

$$Cl_2(aq) + 2I^-(aq) \rightarrow 2Cl^-(aq) + I_2(aq)$$

oxidation states: 0 -1 -1 0

- The oxidation state for chlorine decreases (from 0 to -1), so chlorine is reduced.
- The oxidation state for iodine increases (from -1 to 0), so iodine is oxidised.

Oxidation states in names

> The oxidation state is written in Roman numerals, immediately *after* the element it refers to.

Some compounds contain elements that can exist in more than one oxidation state. When this occurs, the systematic name for the compound includes the oxidation state of the element. For example:

- FeO is called iron(II) oxide
- Fe_2O_3 is called iron(III) oxide.

The names of **oxyanions** should also include an oxidation state. For example:

Ion	Name	Oxidation state of chlorine
ClO^-	chlorate(I)	+1
ClO_2^-	chlorate(III)	+3
ClO_3^-	chlorate(V)	+5
ClO_4^-	chlorate(VII)	+7

> Oxyanions are negative ions that contain oxygen and another element. Their names end with the letters '–ate', e.g. sulfate.

Electron transfer and half-equations

Sodium reacts with chlorine as follows:

$$2Na + Cl_2 \rightarrow 2NaCl$$

This can be written as two separate **half-equations**. From these you can decide which is the oxidation reaction and which is the reduction reaction:

$$2Na \rightarrow 2Na^+ + 2e^-$$

This is oxidation (electron loss). Na changes from oxidation state 0 to +1 and has therefore been oxidised.

$$Cl_2 + 2e^- \rightarrow 2Cl^-$$

This is reduction (electron gain). Cl_2 changes from oxidation state 0 to –1 and has therefore been reduced.

Other redox changes with halide ions

A **displacement reaction** occurs when a more reactive halogen, like Cl_2, is passed into a solution of less reactive halide ions, such as KI.

$$Cl_2(g) + 2I^-(aq) \rightarrow 2Cl^-(aq) + I_2(aq)$$

This reaction can be represented by two half-equations:

$$Cl_2(aq) + 2e^- \rightarrow 2Cl^-(aq) \text{ – this is reduction (electron gain)}$$
$$2I^-(aq) \rightarrow I_2(aq) + 2e^- \text{ – this is oxidation (electron loss)}$$

> The chlorine molecules are acting as the **oxidising agent** (they cause another species to be oxidised and in doing so are reduced themselves). The iodide ions are **reducing agents** (they cause another species to be reduced and in doing so are oxidised themselves).

> K^+ is a spectator ion, so it can be removed from the ionic equation.

QUICK CHECK QUESTIONS

1 Write down the oxidation states of the elements in the following:
 (a) KBr   (b) H_2O (c) Co^{2+}
 (d) PO_4^{3-} (e) MnO_2 (f) $Cr_2O_7^{2-}$
2 Write down the formulae of the following:
 (a) copper(II) chloride   (b) copper(I) oxide
 (c) lead(IV) chloride   (d) manganate(VII) ion
 (e) sodium nitrate(III)
3 Write two half-equations for the following reaction, and identify which equation is the oxidation reaction and which is the reduction reaction.

$$2Ca + O_2 \rightarrow 2CaO$$

4 Zinc is added to copper(II) sulfate solution.
 (a) Write an ionic equation for the reaction.
 (b) Write half-equations for the oxidation and reduction steps in this reaction.

5 Identify the changes in oxidation states in the following reaction:

$$2Br^- + 2H^+ + H_2SO_4 \rightarrow Br_2 + SO_2 + 2H_2O$$

6 Copper reacts with dilute nitric(V) acid to form nitrogen(II) oxide and copper(II) nitrate. The half-equations for the oxidation and reduction reactions are:

$$Cu \rightarrow Cu^{2+} + 2e^-$$
$$NO_3^- + 4H^+ + 3e^- \rightarrow NO + 2H_2O$$

 (a) Construct the ionic equation for the reaction. (*Hint:* you will first need to balance the number of electrons transferred in each half-equation.)
 (b) Show that the overall increase in oxidation state for copper, and decrease in oxidation number for nitrogen, are equal.

The p block: Group 7

Chemical Ideas 11.4 and Chemical Storylines ES3 and ES4

Physical properties

You need to be able to recall the following physical properties of the halogens:

The elements in Group 7 are called the halogens.

	Fluorine (F_2)	Chlorine (Cl_2)	Bromine (Br_2)	Iodine (I_2)
Appearance and state at room temperature	pale yellow gas	green gas	dark red liquid	shiny black solid
Volatility	gas	gas	liquid quickly forms brown gas on warming	sublimes on warming to give a purple vapour
Solubility in water	reacts with water	slightly soluble to give pale green solution	slightly soluble to give red-brown solution	barely soluble, gives a brown solution
Solubility in organic solvents	soluble	soluble to give a pale green solution	soluble to give a red solution	soluble to give a violet solution

All the halogens exist as elements as *diatomic* non-polar molecules (F_2, Cl_2, Br_2 and I_2). The *intramolecular* bonds are covalent and the *intermolecular* bonds are instantaneous dipole-induced dipole bonds. Fluorine is the most volatile halogen as it has the smallest molecules with the fewest electrons. As the size of the molecule and number of electrons increases, so does the strength of the intermolecular bonds. This explains why the physical state of the halogens changes from gas to liquid to solid as you go down the group.

For more information on intermolecular bonds see pages 46 and 47.

You also need to know and be able to explain the following chemical properties of the halogens:

	Fluorine	Chlorine	Bromine	Iodine
Relative reactivity	most reactive	more reactive than bromine or iodine	more reactive than iodine, less reactive than chlorine	least reactive
Halide ions and silver ions		white precipitate of silver chloride	cream precipitate of silver bromide	yellow precipitate of silver iodide
	General reaction, X^- = halide ion: $Ag^+(aq) + X^-(aq) \rightarrow AgX(s)$			
Displacement reactions	displaces chlorine, bromine and iodine	displaces bromine and iodine	displaces iodine	does not displace bromine, chlorine or fluorine
	$Cl_2(aq) + 2Br^-(aq) \rightarrow 2Cl^-(aq) + Br_2(aq)$ $Cl_2(aq) + 2I^-(aq) \rightarrow 2Cl^-(aq) + I_2(aq)$ $Br_2(aq) + 2I^-(aq) \rightarrow 2Br^-(aq) + I_2(aq)$			
Redox reactions	Halogens are all reactive. They tend to remove electrons from other elements – they are oxidising agents. $X_2 + 2e^- \rightarrow 2X^-$ Halogen oxidation state 0 is reduced to oxidation state −1.			

Redox reactions involving halogens

When a solution containing chlorine is added to a solution containing iodide ions, a brown colour appears as iodine is produced:

$$Cl_2(aq) + 2I^-(aq) \rightarrow I_2(aq) + 2Cl^-(aq)$$

Chlorine is a stronger oxidising agent than iodine.

Fluorine is the strongest oxidising agent in Group 7. Fluorine atoms are small and the attraction of a fluorine nucleus for an electron is very strong. As the halogen atoms get bigger, the nucleus is further from the outer shell into which an electron will fit, and the attraction decreases. This is why the reactivity of halogens – and their strength as oxidising agents – decreases down the group.

> Chlorine has been reduced (oxidation state changes from 0 to –1) and iodine has been oxidised (oxidation state changes from –1 to 0).

Electrolysis of solutions containing halide ions

When an electric current is passed through a solution of sodium chloride, chlorine gas bubbles off at the anode. This reaction is the basis of the industrial manufacture of chlorine. Chloride ions lose electrons to the anode and become oxidised:

$$2Cl^- \rightarrow Cl_2 + 2e^-$$

> Similar reactions occur when aqueous solutions of sodium bromide and sodium iodide are electrolysed.

Storage and transport of halogens

- Fluorine is very reactive, in fact, too reactive to store. When needed for chemical reactions, it is made in situ (as it is needed) by electrolysing liquid hydrogen fluoride.
- Chlorine is a highly toxic gas. It is transported by rail or road tanker as a liquid.
- Bromine is transported in lead-lined steel tanks supported in strong metal frames. Transport routes are planned to minimise the risk of accidents, for example, routes are planned to avoid residential areas.

See page 33 for informaton about the benefits and risks of using halogens.

QUICK CHECK QUESTIONS

1 What is the appearance of bromine at room temperature?
2 Describe the appearance of iodine when dissolved in:
   (a) water
   (b) an organic solvent.
3 Explain, in terms of intermolecular bonds, why chlorine is more volatile than bromine.
4 (a) Write the equation for the reaction of potassium iodide solution with silver nitrate.
   (b) What would you see?
5 (a) (i) Write an ionic equation for the reaction that occurs when chlorine gas is bubbled through potassium bromide solution.
       (ii) Describe and explain what you would see.

   (b) If cyclohexane (an organic solvent) was added after the reaction in part (a), what would you see? Explain your answer.
6 Explain why chlorine is a stronger oxidising agent than iodine.
7 (a) Write equations to show the reactions that occur at the anode and cathode when liquid hydrogen fluoride is electrolysed.
   (b) Explain at which electrode oxidation and reduction occur.

43

Electronic structure: sub-shells and orbitals

Chemical Ideas 2.4

- Electrons exist in shells and these are designated $n = 1$, $n = 2$, $n = 3$ etc. The further away a shell is from the nucleus the larger its n number.
- These shells are sub-divided into **sub-shells** labelled **s**, **p**, **d** and **f**.
- Each sub-shell is further divided into **atomic orbitals** – each atomic orbital can hold a maximum of two electrons. These two electrons *must* have opposite (or paired) **spins**.

'Arrows in boxes' are a good way of representing the filling up of orbitals by electrons.

Distribution of electrons in atomic orbitals

Shell	Description	Sub-shells
first	$n = 1$ has only one sub-shell	s
second	$n = 2$ has two sub-shells	s and p
third	$n = 3$ has three sub-shells	s, p and d
fourth	$n = 4$ has four sub-shells	s, p, d and f

- An s sub-shell has 1 orbital holding a maximum of 2 electrons.
- A p sub-shell has 3 orbitals holding a maximum of 6 electrons.
- A d sub-shell has 5 orbitals holding a maximum of 10 electrons.

The arrangement of electrons in shells and orbitals is called the **electronic configuration**.

> Note the 4s orbital is at a lower energy level than the 3d and so fills up before the 3d.

> The arrows represent electrons and the box represents an orbital. Arrows pointing in opposite directions have paired spins.

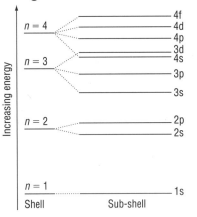

1. The orbitals are filled in order of increasing energy.
2. Where there is more than one orbital at the same energy, the orbitals are first occupied singly by electrons. When each orbital is singly occupied, then electrons pair up in orbitals.
3. Electrons in singly occupied orbitals have parallel spins.
4. Electrons in doubly occupied orbitals have opposite spins.

Representing electron distribution

Each orbital can be represented as a box, and each electron as an arrow. The electronic configurations for nitrogen and sodium can be represented as follows:

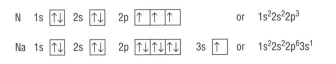

> You may be asked to draw diagrams for any of the elements up to and including krypton (atomic number 36). It would be a good idea to practise these.

Deducing electronic configurations

Using the energy level diagram on the previous page, the electron configuration of an element can be deduced.

WORKED EXAMPLE

STEP 1 Magnesium has an atomic number of 12, and therefore has 12 protons.

STEP 2 An atom of magnesium has 12 electrons.
It has: 2 in shell $n = 1$; 8 in shell $n = 2$; and 2 in shell $n = 3$.

STEP 3 These are further divided into sub-shells written as $1s^2 2s^2 2p^6 3s^2$.
The notation for phosphorus (atomic number 15) is 2.8.5 in shells, and $1s^2 2s^2 2p^6 3s^2 3p^3$ in sub-shells.
Potassium is 2.8.8.1 or $1s^2 2s^2 2p^6 3s^2 3p^6 4s^1$.

> The abbreviated electronic configuration of potassium is [Ar] $4s^1$, where [Ar] represents the electronic configuration of argon.

s, p and d blocks

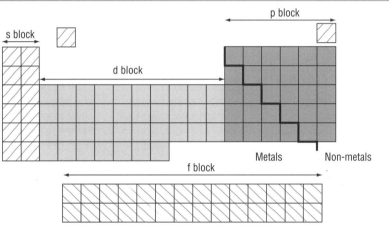

- Group 1 and 2 elements all have one or two electrons, respectively, in their outermost sub-shell, which is an s orbital. These are known as the s-block elements. For example, magnesium is $1s^2 2s^2 2p^6 \mathbf{3s^2}$

- Group 3, 4, 5, 6, 7 and 0 elements all have three, four, five, six, seven or eight electrons, respectively, in their outermost sub-shell, which are p orbitals. They are known as the p-block elements. For example, selenium is $1s^2 2s^2 2p^6 3s^2 3p^6 4s^2 3d^{10} \mathbf{4p^4}$

- The first transition block elements have electrons which are filling the d sub-shell. These are known as the d-block elements. For example, cobalt is $1s^2 2s^2 2p^6 3s^2 3p^6 4s^2 \mathbf{3d^7}$

- Further transition metals have electrons which are filling the f sub-shell. These are known as the f-block elements.

> There is a Periodic Table inside the front cover of this revision guide.

QUICK CHECK QUESTIONS

1 **(a)** How many sub-shells does the third shell have?
 (b) Name them.
2 What is the electronic configuration for sulfur, atomic number 16?
3 Which elements have these electronic configurations:
 (a) $1s^2 2s^2 2p^5$
 (b) $1s^2 2s^2 2p^6 3s^2 3p^6 4s^2$
 (c) $1s^2 2s^2 2p^6 3s^2 3p^6 4s^2 3d^3$?
4 Classify the elements in Question 3 as s-, p- or d-block elements.

5 Write the electronic configuration for the following elements:
 (a) titanium (22 electrons)
 (b) aluminium (13 electrons)
 (c) bromine (35 electrons).
6 Identify the following elements from the electronic configuration of their outermost shell:
 (a) element X: $6s^1$
 (b) element Y: $4s^2 4p^1$
 (c) element Z: $5s^2 5p^6$.

Bonds between molecules: temporary and permanent dipoles

Chemical Ideas 5.3 and 3.1

In any liquid or solid there are bonds *between* molecules. These are called **intermolecular bonds**.

Polar molecules

A **dipole** occurs when a molecule (or part of a molecule) has a positive end and a negative end.

When a molecule has a dipole, we say it is **polarised**. Molecules with a *permanent* dipole are **polar molecules**.

For example:

$\delta+ \quad \delta-$
H—Br

Temporary, or instantaneous, dipoles

If a molecule does not have a permanent dipole, the electron density in the molecule may be unevenly distributed at any one time – it has an **instantaneous dipole**. The swirling electron density distribution changes and so the polarity will change.

Electron cloud evenly distributed; no dipole

At some instant, more of the electron cloud happens to be at one end of the molecule than the other; molecule has an instantaneous dipole

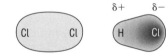

An unpolarised Cl_2 molecule finds itself next to an HCl molecule with a permanent dipole

Electrons get attracted to the positive end of the HCl dipole, inducing a dipole in the Cl_2 molecule

If other molecules are close to a molecule with a dipole these may cause an effect and produce an **induced dipole**.

Intermolecular bonds

Remember – intermolecular bonds need to be broken in order for a substance to melt or boil. The stronger these intermolecular bonds are then the more energy is needed to break them, so the higher the melting or boiling point.

Molecular substances which contain dipoles attract each other. Two of these kinds of attraction are instantaneous dipole–induced dipole bonds and permanent dipole–permanent dipole bonds.

1. Instantaneous dipole–induced dipole bonds

These are the weakest type of intermolecular bond. They can happen in *all types of molecule* – even those with a permanent dipole already.

Consider krypton atoms. Their electrons are continually moving, creating instantaneous dipoles. When other krypton atoms approach an atom with an instantaneous dipole they will produce an induced dipole. This is an **instantaneous dipole–induced dipole** attraction. Because the electron cloud in the krypton atoms is continually moving, these instantaneous dipole–induced dipole interactions are continually forming and breaking. The more electrons an atom or molecule has then the greater these attractions and the higher the boiling point of the substance.

The only intermolecular bonds between chains of poly(ethene) are instantaneous dipole–induced dipoles and yet poly(ethene) is solid at room temperature. This is because the chains are long and can pack closely together. Although the intermolecular bonds are very weak, there are a lot of them.

2. Permanent dipole–permanent dipole bonds

Molecules with permanent dipoles have atoms with different **electronegativity** values. The slightly positively charged end of a molecule attracts the slightly negatively charged end of the next molecule, and an intermolecular bond occurs. This is stronger than the instantaneous dipole–induced dipole bonds between noble gas atoms. Since the dipoles in both molecules are permanent, this type of intermolecular attraction is called a **permanent dipole–permanent dipole** bond.

Permanent dipole–permanent dipole bonds are stronger than instantaneous dipole–induced dipole bonds but are weaker than hydrogen bonds (see pages 70 and 71).

Permanent dipole–permanent dipole bonds are responsible for holding polyester molecules together.

> You may be asked to identify the strongest intermolecular bond in a given molecule and label it on a diagram. Always label the bond polarities and identify the attraction between the polarised bonds, as in the diagrams here.

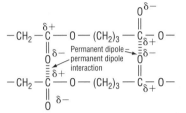

$$\delta+ \quad \delta- \quad \delta+ \quad \delta- \quad \delta+ \quad \delta-$$
$$\text{---H} - \text{Br} \text{---H} - \text{Br} \text{---H} - \text{Br}$$

Permanent dipole–permanent dipole bonds between HBr molecules.

Section of two polyester chains.

Electronegativity

The degree to which an atom of an element attracts electrons is called its **electronegativity**. The more electronegative an element is then the greater its attraction for electrons. The order of electronegativity values for some common elements is:

F > O > Cl > Br and N > I > S > C > H

The difference between the electronegativities of C and H is so small that bonds between them can be considered to be non-polar.

> Electronegativity generally increases towards the top and right of the Periodic Table.

Bond polarity and polar molecules

A polar molecule is one that has a permanent dipole, for example ethanoic acid:

If the differences in electronegativities of the elements in a molecule are very small, the dipole is negligible – for example in CH_4. Sometimes even if the bonds are polar a molecule might not have a dipole. This is due to the arrangement of the polar bonds in the molecule. For example, in tetrachloromethane each C–Cl bond is polar but the symmetrical arrangement of the chlorine atoms means that there is no overall dipole.

Tetrachloromethane has no overall molecular dipole.

> To understand work on molecular dipoles, you need to understand about molecular shape. You can find out about these ideas on page 10.

1 Would you expect xenon or krypton to have the higher boiling point? Explain your answer.
2 Why do branched-chain hydrocarbons have lower boiling points than straight-chain hydrocarbons of the same relative formula mass?
3 Explain how instantaneous dipole–induced dipole bonds arise between molecules of hydrogen, H_2.
4 What type(s) of intermolecular bonds would you expect between HCl molecules? Explain your answer.

5 What is the strongest type of intermolecular bond between molecules of:
   (a) ethane, C_2H_6
   (b) trichloromethane, $CHCl_3$
   (c) hydrogen and hydrogen bromide, H_2/HBr?
6 Draw a diagram to show the intermolecular bonding between two poly(chloroethene) chains.
7 Is carbon dioxide a polar molecule? Explain your answer.

Halogenoalkanes

Chemical Ideas 13.1

Halogenoalkanes are sometimes called haloalkanes.

The homologous series of the halogenoalkanes is an alkane series with hydrogen atoms substituted by one or more halogen atoms. They are often shown as R–Hal, where Hal could be F, Cl, Br or I.

Naming halogenoalkanes

The alkane chain name is *prefixed* with the name of the halogen. The halogens are listed in alphabetical order, with a number indicating the position of each.

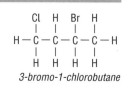

3-bromo-1-chlorobutane

Each halogen atom is prefixed with a number – e.g. 1,1-dichloro-2-iodopropane.

```
     Cl  I   H
     |   |   |
Cl — C — C — C — H
     |   |   |
     H   H   H
```

There are two chlorine atoms attached to carbon 1 so there are two numbers here.

Physical properties of halogenoalkanes

The boiling points increase with a heavier halogen atom (R–I > R–F) or with increasing numbers of halogen atoms (CCl_4 > CH_2Cl_2). As the halogen introduced is larger or the number of halogen atoms increases, the overall number of electrons increases. This increases the instantaneous dipole–induced dipole bonds (see pages 46 and 47). With stronger intermolecular bonds, more energy is needed to pull the molecules apart from each other, so the boiling point is higher.

Bond enthalpies and reactivity of halogenoalkanes

Bond within the molecule	Bond strength	Reactivity
C–F	decreasing strength	increasing reactivity
C–Cl		
C–Br		
C–I		

The C–Hal bond becomes weaker as the size of the halogen atom increases. This makes the bond easier to break and the compounds become more reactive. Although the C–F bond is the most polar, fluoroalkanes are very unreactive. This shows that it is bond strength rather than bond polarity that has the greatest effect on the reactivity of halogenoalkanes.

- Fluoro- compounds are very unreactive.
- Chloro- compounds are reasonably stable in the troposphere and can react to produce chlorine radicals that deplete ozone.
- Bromo- and iodo- compounds are reactive and so are useful as intermediates in chemical synthesis.

Reactions of halogenoalkanes

Conditions: gas phase with high temperatures; or the presence of UV radiation (e.g. in the stratosphere).

1 Homolytic fission (forming radicals)

$$R \frown Cl \xrightarrow{h\nu} R^\bullet + Cl^\bullet$$

2 Heterolytic fission

Conditions: dissolved in a polar solvent such as an ethanol/water mixture.

$$R-Hal \longrightarrow R^+ + Hal^-$$

The carbon–halogen bond breaks to give ions. If the polar C–Hal bond is broken completely a negative halide ion moves away, leaving the C group positively charged. This is now a **carbocation**.

$$CH_3-\underset{\underset{CH_3}{|}}{\overset{\overset{CH_3}{|}}{C}}-Cl \longrightarrow CH_3-\underset{\underset{CH_3}{|}}{\overset{\overset{CH_3}{|}}{C}}{}^+ + Cl^-$$

2-chloro-2-methylpropane   carbocation   *chloride* ion

3 Substitution reactions

$$R–Hal + X^- \rightarrow R–X + Hal^-$$

In this case the C–Hal bond breaks and the halogen atom is replaced by another functional group. Since the halogen is replaced by a nucleophile these reactions are called **nucleophilic substitution** reactions.

> X^- is an example of a nucleophile.

> A full headed arrow indicates the movement of a pair of electrons.
> A half-headed arrow indicates the movement of one electron.

WORKED EXAMPLE

The mechanism for the nucleophilic substitution reactions of halogenoalkanes.

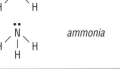

STEP 1 The δ+ carbon from the carbon–halogen bond is attacked by the nucleophile.
STEP 2 The lone pair of electrons on the nucleophile forms a new bond with the carbon.
STEP 3 At the same time the carbon–halogen bond breaks, giving a halide ion.

> Nucleophiles have one or more lone pairs of electrons that they can donate to form new bonds. Examples include:
>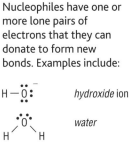
> $H–O$ hydroxide ion
> water
> ammonia

Nucleophile	Equation	Product	Reaction conditions
HOH	$R–Hal + H_2O \rightarrow R–OH + H^+ + Hal^-$	alcohol	heat under reflux: this is sometimes called **hydrolysis**
OH^-	$R–Hal + OH^- \rightarrow R–OH + Hal^-$	alcohol	heated under reflux with NaOH(aq) with ethanol as solvent
NH_3	$R–Hal + NH_3 \rightarrow R–NH_2 + Hal^- + H^+$	amine	the halogenoalkane is heated with concentrated ammonia solution in a sealed tube

> You will need to learn the reaction conditions for these reactions.

Preparation of halogenoalkanes

You need to know the practical details for the preparation and purification of a halogenoalkane (see **Activity ES6.3** for details).

A tertiary alcohol like 2-methylpropan-2-ol will react at room temperature with concentrated hydrochloric acid to form 2-chloro-2-methylpropane. The reaction is carried out in a separating funnel. The chloroalkane is immiscible in water and forms a layer above the aqueous products. There are five stages in the purification of the chloroakane:

- the upper layer containing the chloroalkane is run off into a clean beaker
- the chloroalkane is shaken with a solution of sodium hydrogencarbonate to remove any acidic impurities
- the chloroalkane layer is run off for a second time
- anhydrous sodium sulfate (a drying agent) is added to remove any water
- the chloroalkane is purified by distillation.

> Check densities, it could be that the halogenoalkane you are preparing has a greater density than water. If this is the case it will in fact be the lower layer.

The mechanism is as follows:

QUICK CHECK QUESTIONS

1 Draw the skeletal formula for $CH_3CH_2CHBrCCl_2CHICH_3$ and name the compound.

2 Which in each of the following pairs of molecules has the higher boiling point? Explain your reasoning.
  (a) CH_3Br and CH_3F
  (b) CH_3Br and CBr_4

3 Why is iodoethane more reactive than fluoroethane?

4 Explain the term *nucleophile* and give two examples.

5 What are the conditions needed for ammonia to react with a halogenoalkane?

6 Draw the mechanism for the nucleophilic substitution reaction between the hydroxide ion and 1-iodopropane.

7 What are the five main stages in the purification of 1-bromobutane after its production from butan-1-ol?

8 What happens when a halogenoalkane undergoes *hydrolysis*?

Greener industry

Chemical Ideas 15.1–15.7

This section looks at aspects of a chemical manufacturing process and the rapidly developing area of 'green chemistry', which involves designing chemical processes that reduce or eliminate waste and the use of hazardous materials.

Batch process or continuous process?

A **chemical plant** is the name given to the place where chemicals are manufactured. The term 'plant' is also used to describe the site equipment such as reaction vessels, storage facilities and pipework. The diagram shows a typical chemical process.

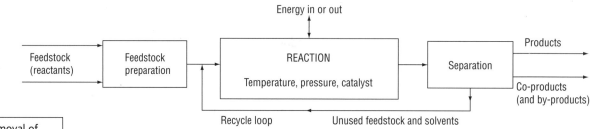

When chemicals are manufactured using a **batch** process, the reactants are placed in a reaction vessel and allowed to react. Once the reaction is over, the products are removed and the vessel is cleaned and made ready for the next batch.

In a **continuous** process, the starting materials are regularly fed in at one end of the plant and the product emerges at the other end.

Batch process	Continuous process
Advantages: • cost effective for small quantities • low capital costs for plant • different products can be made using the same vessel *Disadvantages*: • filling, emptying and cleaning reaction vessel is time-consuming • may require larger workforce • contamination possible if making different products	*Advantages*: • suited to high-tonnage products • can operate for months at a time without a shut-down • more easily automated, so workforce can be smaller *Disadvantages*: • less flexible – usually designed to make one product • much higher capital costs in building the plant • not cost effective if run below full capacity

Raw materials

Raw materials are the materials that feedstocks are prepared from. They are usually obtained from the ground (e.g. rock salt, limestone or crude oil) or from the atmosphere (e.g. air). Raw materials are processed and converted into **feedstocks** – the reactants which are fed in at the start of the process.

For example, in the electrolysis of brine:

raw material ⟶ feedstock ⟶ products

rock salt   brine   chlorine, sodium hydroxide
           (sodium chloride solution)   and hydrogen

Input or removal of energy may be required at any stage

Dyes, pharmaceuticals and pesticides are made using a batch process.

Ethene, ammonia and sulfuric acid are made using a continuous process.

The main construction material for a chemical plant is mild steel, but where corrosion resistance is required then more expensive materials are used – e.g. glass-lined steel, specialist alloys or reinforced plastics.

Preparing the feedstock is often a large part of any manufacturing process.

In the example on page 50, chlorine is the main desired product. Sodium hydroxide and hydrogen are **co-products** – i.e. they are produced at the same time as the desired product via the same reaction. As the amount of desired product increases so do the co-products. In this case, they are all useful chemicals that can be sold to generate further profits for the manufacturer.

Many reactions produce **by-products** – these are the result of unwanted side reactions. The conditions of any chemical process are designed to *increase* the amount of desired product and *decrease* the amount of by-products as much as possible.

Costs and efficiency

The amount of profit a chemical company makes depends on income from sales of product and on costs. **Fixed costs** are the same, irrespective of how much product is made. **Variable costs** depend on the amount of product manufactured.

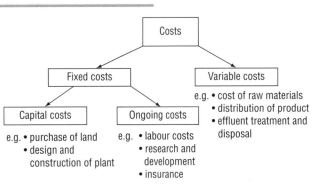

The **efficiency** of a chemical process depends on many factors, including the operating temperature and pressure. High temperatures increase the rate of a reaction but may affect the position of equilibrium adversely and reduce yields. Operating at high pressure can improve the yield for *some* reactions but requires more expensive reaction vessels and has safety implications. **Recycling** unreacted feedstock is an important way of reducing costs.

Efficient use of **energy** is also essential. Energy costs (for electricity, gas or fuel oil) are high and so any heat produced during an exothermic reaction is converted, using a heat exchanger, into hot water or steam for use in other parts of the plant.

Plant location

Chemical plants were traditionally sited near sources of raw materials. Nowadays, factors which affect the location of a chemical plant include availability of:

- A good transport network – a sea terminal and good rail and road links are all important to bring in raw materials and distribute products.
- Labour – new chemical plants are often built near existing ones because of the availability of skilled labour.
- Shared facilities – the product from one plant may be the feedstock for another.
- Cheap energy – many chemical industries use vast amounts of energy.
- Water – this can be important for manufacture of high-tonnage materials.

> The site of *Anglesey Aluminium* is near a deep-water sea terminal to import alumina (feedstock) and a power station that can provide cheap off-peak electricity.

Health and safety

Safety is a major consideration in all operations in the chemical industry, and UK companies have to abide by national and EU legislation. Regular safety training is provided. Every stage of a manufacturing process is checked to reduce exposure of employees to hazardous chemicals or procedures.

> Typical safety features include eye-baths, emergency showers, toxic gas refuges, breathing apparatus, emergency control rooms and on-site medical centres and fire brigade.

Waste disposal

There are strict limits on the amount of hazardous chemicals that can be released and companies that breach these limits face heavy fines. All chemical waste must now be treated before disposal.

> Ingenious ways are used to deal with potential pollutants. E.g. SO_2 is converted into sulfuric acid, which can be sold to increase profits.

Percentage yield and atom economy

The **percentage yield** for a reaction is calculated using the equation:

$$\% \text{ yield} = \frac{\text{actual mass of product}}{\text{theoretical maximum mass of product}} \times 100$$

The **atom economy** for a reaction is calculated using the equation:

$$\% \text{ atom economy} = \frac{\text{relative formula mass of useful product}}{\text{relative formula mass of reactants used}} \times 100$$

WORKED EXAMPLE

Ethanol can be converted into ethene in the following reaction:

$$C_2H_5OH \rightarrow C_2H_4 + H_2O$$

In one reaction, 23.0 g of ethanol produced 6.0 g of ethene. Calculate the percentage yield and the atom economy of this reaction.

STEP 1 Moles of ethanol used $= \dfrac{\text{mass}}{M_r} = \dfrac{23.0}{46.0} = 0.50$ mole

STEP 2 We can see from the equation that 0.50 mole of ethanol should, in theory, produce 0.50 mole of ethene, or 14.0 g (28.0×0.50).

So % yield $= \dfrac{6.0}{14.0} = 43\%$

STEP 3 % atom economy $= \dfrac{28.0}{46.0} = 61.0\%$

Percentage yield used to be a significant factor in deciding if a chemical process was economically viable. With 'green chemistry' high on the agenda, atom economy is just as important.

QUICK CHECK QUESTIONS

1 Explain the difference between a batch process and a continuous process.
2 Explain the difference between a by-product and a co-product.
3 Suggest if the following processes are batch or continuous:
   (a) catalytic cracking of gas oil in the petrol industry
   (b) fermentation of grapes to make wine.
4 **(a)** What are feedstocks?
   (b) Name the feedstocks used for the manufacture of ammonia in the Haber process.
   (c) Name the raw materials used in the Haber process.
5 Suggest why locating a new chemical plant on an existing site of chemical manufacture can lower costs.
6 Why might a company manufacturing bleach want to site itself near a chlor–alkali plant?
7 Epoxyethane, an important feedstock in the chemical industry, is made by the following reaction

ethene + ½ O₂ → epoxyethane (Ag, 300 °C)

(a) What is the atom economy of the reaction?
(b) In one reaction, 5.6 kg of ethene produced 1.32 kg of epoxyethane. Calculate the percentage yield for this reaction.
(c) Suggest what happens to the unreacted ethene in the reaction.
8 Chloroethene, the monomer for poly(chloroethene) manufacture, can be produced in a two-step process

Step 1: $CH_2{=}CH_2 + 2HCl + \frac{1}{2}O_2 \xrightarrow[250\,°C]{CuCl_2} CH_2ClCH_2Cl + H_2O$
ethene → 1,2-dichloroethane

Step 2: $CH_2ClCH_2Cl \xrightarrow{500\,°C} CH_2{=}CHCl + HCl$
chloroethene

(a) Calculate the atom economies for step 1 and step 2.
(b) What could be done to improve the efficiency of the process overall?
(c) 420 g of ethene reacted with excess hydrogen chloride and oxygen to make 337.5 g of chloroethene. What was the percentage yield in this reaction?

The Atmosphere (A)

This module looks at the chemical and physical processes that go on in the atmosphere. It concentrates firstly on looking at the chemistry responsible for the depletion of the ozone layer, and secondly on global warming. As you work through this module you will cover the following concepts. 'CI' refers to sections in your *Chemical Ideas* textbook. 'Storylines' refers to your *Chemical Storylines* textbook.

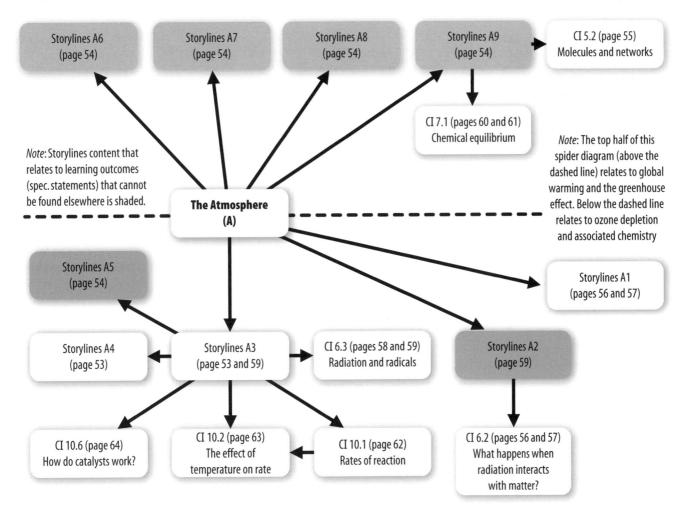

Storylines A6 (page 54)

Storylines A7 (page 54)

Storylines A8 (page 54)

Storylines A9 (page 54)

CI 5.2 (page 55) Molecules and networks

CI 7.1 (pages 60 and 61) Chemical equilibrium

Note: Storylines content that relates to learning outcomes (spec. statements) that cannot be found elsewhere is shaded.

The Atmosphere (A)

Note: The top half of this spider diagram (above the dashed line) relates to global warming and the greenhouse effect. Below the dashed line relates to ozone depletion and associated chemistry

Storylines A5 (page 54)

Storylines A1 (pages 56 and 57)

Storylines A4 (page 53)

Storylines A3 (page 53 and 59)

CI 6.3 (pages 58 and 59) Radiation and radicals

Storylines A2 (page 59)

CI 10.6 (page 64) How do catalysts work?

CI 10.2 (page 63) The effect of temperature on rate

CI 10.1 (page 62) Rates of reaction

CI 6.2 (pages 56 and 57) What happens when radiation interacts with matter?

From *Chemical Storylines* you will need to be aware of the following:

Chlorofluorocarbons (CFCs) (Storylines A3 and A4)

CFCs are used as refrigerants, propellants in aerosols, blowing agents for making expanded plastics and as cleaning solvents. CFCs are good at these jobs because of their low reactivities, low boiling points, low toxicity and high stability.

It is this last property that causes the problem. CFCs are estimated to have a lifetime in the troposphere of approximately 100 years. When they reach the stratosphere they undergo photodissociation to produce chlorine radicals, which then go on to remove ozone.

In the early 1970s Sherry Rowland and Mario Molina predicted that CFCs would damage the ozone layer. They were proved correct in 1985 when a team, led by Joe Farman, identified a hole in the ozone layer over the Antarctic using ultraviolet spectroscopy. The computers for NASA satellites had treated the very low ozone readings as anomalies. Conclusive proof that CFCs were responsible for ozone depletion was

See page 59 for more information.

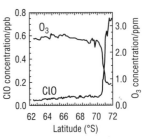

obtained when an aircraft flew through the stratosphere over Antarctica and measured the concentrations of ClO and O_3. The rapid fall in O_3 occurred at exactly the point where ClO concentrations rose. The link had finally been made. Findings from many parts of the scientific community had succeeded in validating the results of all the work.

Alternatives to CFCs (Storylines A5)

Replacements have been used for CFCs, although they in turn have their own problems.

Replacement	Advantages	Disadvantages
hydrochlorofluorocarbons (HCFCs) and hydrofluorocarbons (HFCs)	H–C bonds are broken down in the troposphere before the compounds have chance to reach the stratosphere	are greenhouse gases that contribute to global warming
alkanes	alkanes don't contain chlorine	are flammable and are greenhouse gases

The greenhouse effect (Storylines A6–A9)

> The greenhouse effect keeps the troposphere at a temperature that enables life to exist on Earth.

High-energy radiation from the Sun (mostly visible and UV radiation) reaches the Earth's surface and some frequencies are absorbed. The Earth's surface is warmed and re-emits lower energy infrared radiation. Greenhouse gases, such as CH_4 and CO_2, absorb some of this radiation – the rest escapes into space. A steady state is reached where Earth radiates and absorbs energy at the same rate. The absorption of infrared radiation by 'greenhouse gases' can cause atmospheric warming in two ways:

- Some infrared radiation is re-emitted by the molecules *in all directions* – some energy is radiated back towards Earth and some out into space.
- Absorption of infrared radiation increases the vibrational energy of the molecules – their bonds vibrate *more vigorously*. Energy is transferred to other molecules in the atmosphere by collisions. This increases their kinetic energy and raises the average temperature of the atmosphere

Carbon dioxide and water are important greenhouse gases. CO_2 and H_2O absorb in two bands across the Earth's radiation spectrum. Between these two bands is a 'window' where infrared radiation can escape without being absorbed. About 70% escapes through this fixed range of frequencies.

Is human activity responsible for global warming? (Storylines A7–A9)

> The global warming potential depends on how efficiently a gas absorbs infrared radiation and on its atmospheric lifetime. CO_2 has a GWP of 1, and CH_4 has a GWP of 25. CFC-11 has a GWP of 4750.

Human activities *are* increasing the atmospheric concentrations of 'natural' greenhouse gases (e.g. CO_2) and gases that are not naturally present (e.g. CFCs). Although only present in small amounts, each CFC molecule has a large 'global warming potential' (GWP).

These gases absorb radiation in the 'window' through which energy normally escapes into space. This leads to an **enhanced greenhouse effect**. The majority of scientists now believe that human activities are contributing to global warming.

In order to reduce global warming, the amount of CO_2 put into the atmosphere needs to be reduced. This could be done by:
- reducing our consumption of fossil fuels
- using alternative energy sources (wind, solar, tidal, nuclear)
- increasing photosynthesis
- burying or reacting carbon dioxide. (**Activity A9.3**)

Molecules and networks

Chemical Ideas 5.2

Some elements and compounds form giant structures, covalently bonded together. We call these **network** or **giant structures.** Two examples are diamond and silicon(IV) oxide.

Diamond

Diamond is made up of carbon atoms.

- Each carbon atom is joined tetrahedrally to four other carbon atoms, by strong covalent bonds.
- The very strong C–C bonds and highly symmetrical network structure make diamond the hardest naturally occurring substance.

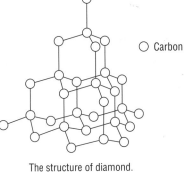

○ Carbon

The structure of diamond.

Silicon(IV) oxide

Silicon atoms form four bonds.

- Silicon bonds covalently to four oxygen atoms.
- Quartz is a pure form of silicon(IV) oxide. It is an extended network of SiO_4 units. Each silicon atom has a half-share in four oxygen atoms.

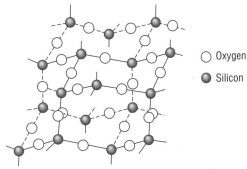

○ Oxygen

● Silicon

The structure of silicon(IV) oxide (SiO_2).

Differences between CO_2 and SiO_2

Carbon dioxide has a **molecular** structure with three atoms bonded in a linear arrangement, O=C=O. There are weak intermolecular bonds (the bonds *between* molecules). Little energy is needed to pull the molecules apart, so solid carbon dioxide sublimes at a low temperature. The intramolecular bonds in carbon dioxide are polar, so it dissolves in water easily.

Silicon atoms are larger than carbon atoms and have more electrons. They are unable to make double bonds, therefore discrete SiO_2 molecules are not formed. Silicon(IV) oxide has a giant network structure. It requires a lot of energy to overcome these intramolecular bonds (bonds *within* the network). Consequently, silicon(IV) oxide has a high melting point and a high boiling point and does not dissolve in water.

> $\delta-\ \ \delta+\ \delta-$
> O=C=O
>
> Polarity of bonds in carbon dioxide.

> Solid CO_2 sublimes at $-78\,°C$ (195 K). SiO_2 melts at $1610\,°C$ (1883 K) and boils at $2230\,°C$ (2503 K).

QUICK CHECK QUESTIONS

1   Why is diamond the hardest naturally occurring substance?
2   Why is CO_2 a gas at room temperature, while SiO_2 is a solid?
3   Why is CO_2 soluble in water, while SiO_2 is insoluble?
4   What types of intermolecular bonds exist between CO_2 molecules?

5   An element in Group 6 of the Periodic Table has a melting point of 113 °C and a boiling point of 445 °C. Would you expect this element to have a network structure or a molecular structure?

What happens when radiation interacts with matter?

Chemical Ideas 6.2

The atmosphere

Increasing distance from the Earth's surface ↑

ionosphere
stratosphere
troposphere

The atmosphere is divided into three sections – the troposphere, the stratosphere and the ionosphere. The troposphere, the section of the atmosphere closest to the surface of the Earth, is made up of approximately 78% nitrogen and 21% oxygen. The remaining 1% is a mixture of gases, mainly argon (0.93%) and carbon dioxide (0.038%).

Human activity alters the proportion of some of the naturally occurring gases, which then become major pollutants – e.g. carbon dioxide. Other pollutants, such as chlorofluorocarbons (CFCs), occur only as a result of human activity.

In order to convert ppm to % divide by 10 000 – e.g. 500 ppm = 0.05%.
In order to convert % to ppm multiply by 10 000 – e.g. 2% = 20 000 ppm.

The concentration of the major gases is usually quoted in per cent by volume. Those present at less than 1% by volume are often quoted in parts per million (ppm).

Energy interacts with matter

These energies are all **quantised** – they exist at certain *fixed* levels. It is possible to bring about changes in any of these energies – i.e. move them from one fixed level to another. These energy changes are caused by radiation interacting with matter at specific frequencies.

Any molecule will have certain energies associated with its behaviour. In increasing order of energy these include:

- translational energy associated with the molecule moving around as a whole
- rotational energy associated with the molecule rotating as a whole
- vibrational energy associated with the vibration of bonds within the molecule
- electronic energy associated with electrons moving from one level to another.

Energy changes and the electromagnetic spectrum

	radiofrequency		microwave	infrared	visible	UV	X-rays	γ-rays

Frequency (Hz)	10^5	10^6	10^7	10^8	10^9	10^{10}	10^{11}	10^{12}	10^{13}	10^{14}	10^{15}	10^{16}	10^{17}	10^{18}	10^{19} 10^{20}
Wavelength (m)		10^3			1			10^{-3}			10^{-6}				10^{-9}	

Different types of electromagnetic radiation have photons of different energy associated with them. When the electromagnetic radiation interacts with matter it will bring about changes in the energy associated with that matter. The type of energy changes depend on the type of radiation absorbed.

increasing energy ↑	Type of change in energy	Type of radiation absorbed
	electronic	ultraviolet or visible
	vibrational	infrared
	rotational	microwave

E is the energy associated with one photon (in joules); v is the frequency of the radiation (in Hz); h is the Planck constant, 6.63×10^{-34} J Hz^{-1}

The energy of photons absorbed to bring about these changes is related to the frequency of the radiation by the equation:

$$E = hv$$

WORKED EXAMPLE

What is the energy of a photon of blue light with a frequency of 7×10^{14} Hz?

$$E = h\nu = 6.63 \times 10^{-34} \times 7 \times 10^{14} = 4.6 \times 10^{-19} \, \text{J}$$

> Remember the units; this example gives an answer in joules.

WORKED EXAMPLE

Bromomethane (CH_3Br) can reach the stratosphere where it undergoes photodissociation.

(a) Calculate the energy required to break a C–Br bond in *one molecule* of bromomethane. The bond enthalpy of the C–Br bond is +276 kJ mol^{-1}

$$E = \frac{276 \times 1000}{6.02 \times 10^{23}} = 4.58 \times 10^{-19} \, \text{J}$$

> Remember: Convert kJ to J by multiplying by 1000

> 6.02×10^{23} is the Avogadro constant – the number of particles in one mole

(b) Calculate the frequency of radiation corresponding to this energy.

$$\nu = \frac{E}{h} = \frac{4.58 \times 10^{-19}}{6.63 \times 10^{-34}} = 6.9 \times 10^{14} \, \text{Hz}$$

> This corresponds to UV radiation

(c) Use the equation $c = \nu\lambda$ to calculate the wavelength of this radiation. ($c = 3.00 \times 10^8 \, \text{m s}^{-1}$)

Rearrange $c = \nu\lambda$

$$\lambda = \frac{c}{\nu} = \frac{3.00 \times 10^8}{6.9 \times 10^{14}} = 4.34 \times 10^{-7} \, \text{m}$$

Electronic changes when molecules absorb radiation

Electrons in molecules occupy definite energy levels. One of three things can happen when a molecule absorbs visible or UV radiation:

- electrons can be excited to a higher electronic energy level; the electrons will return to their original energy levels in time, releasing the energy that has been absorbed

- chemical bonds can break and radicals form – this type of bond dissociation is called photodissociation

- an electron is ejected from a molecule which then becomes ionised.

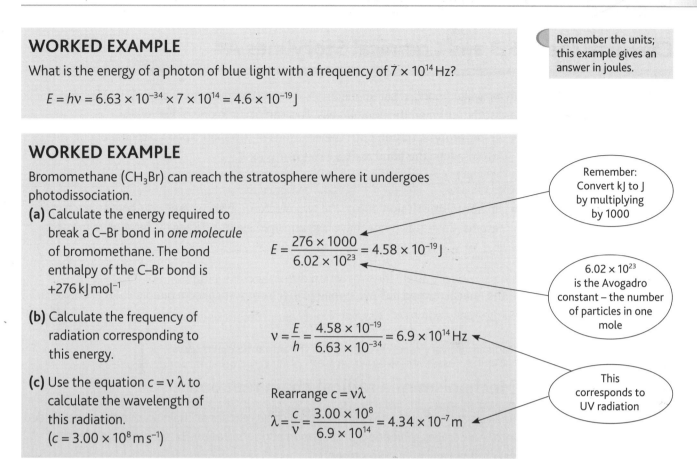

QUICK CHECK QUESTIONS

1 (a) There is 383 ppm of CO_2 in the atmosphere. Convert 383 ppm into a percentage.
  (b) There is 0.000 05% of hydrogen in the atmosphere. Convert this into ppm.

2 (a) What is the energy of a photon of frequency 3.8×10^{13} Hz?
  (b) To which type of radiation does this correspond?
  (c) The change between which type of energy levels is brought about by this radiation?

3 (a) Explain what is meant by the word 'quantised'.
  (b) State three things that can happen when a molecule absorbs a photon of UV radiation.

4 Dioxygen molecules (O_2) undergo photodissociation in the stratosphere.
  (a) The bond enthalpy of the O=O bond is +498 kJ mol^{-1}. Calculate the energy required to break the O=O bond in *one* dioxygen molecule.
  (b) Calculate the frequency of radiation corresponding to this energy.

Radiation and radicals

Chemical Ideas 6.3 and Chemical Storylines A4

⌢↗ indicates the movement of two electrons while ⌢↷ indicates the movement of one electron.

In a covalent bond, a pair of electrons is shared between two atoms. The electron pair is redistributed when the bond breaks. This can occur in one of two ways:

- In **heterolytic fission** both the electrons of the shared pair go to just one of the atoms when the bond breaks – this forms ions.

$$H \overset{\frown}{:} Cl \longrightarrow H^+ + Cl^-$$

*A **radical** is a species with one or more unpaired electrons.*

- In **homolytic fission** one of the two electrons in the shared pair goes to each of the atoms. Both atoms now have one unpaired electron. Radicals have been formed.

$$Br \overset{\frown}{:} Br \longrightarrow 2Br\,\bullet$$

If the radical formed has *two* unpaired electrons it is called a biradical. An example is O_2.

$$\bullet O - O \bullet$$

Radicals are very reactive because of the unpaired electrons.

Mechanism of a radical chain reaction

Radical reactions are fast, are often initiated by heat or light and normally occur in the gas phase.

Radicals are highly reactive and undergo chain reactions. Chain reactions can be divided into three stages – *initiation*, *propagation* and *termination*. Be sure you can identify which stage an equation represents:

- **Initiation** – there are no radicals at the beginning of this stage but radicals are formed by the end of the stage.

In this equation $h\nu$ represents energy.

$$Cl \overset{\frown}{:} Cl \xrightarrow{h\nu} 2Cl\bullet$$

- **Propagation** – there are radicals at the start of this stage, and new radicals are formed by the end of the stage.

$$Cl\bullet + \overset{\frown}{H} \overset{\frown}{:} H \longrightarrow Cl - H + H\bullet$$

$$H\bullet + Cl \overset{\frown}{:} Cl \longrightarrow H - Cl + Cl\bullet$$

- **Termination** – the reaction is terminated when two radicals collide. For example

$$H\bullet + H\bullet \longrightarrow H - H$$

The reactions of alkanes with halogens

A halogen can substitute a hydrogen in an alkane chain, via a **radical substitution** mechanism, to produce a halogenoalkane:

- **Initiation** Homolytic fission occurs in the presence of UV light:

When radicals form due to the presence of light, this is known as photodissociation.

$$Cl_2 \xrightarrow{h\nu} 2Cl\bullet$$

- **Propagation**

$$CH_4 + Cl\bullet \longrightarrow CH_3\bullet + HCl$$

$$CH_3\bullet + Cl_2 \longrightarrow CH_3Cl + Cl\bullet$$

- **Termination**

$$CH_3\bullet + CH_3\bullet \longrightarrow C_2H_6$$

$$CH_3\bullet + Cl\bullet \longrightarrow CH_3Cl$$

Formation and destruction of ozone (Storylines A2 and A3)

In the stratosphere, dioxygen molecules (O_2) can absorb ultraviolet radiation of the right frequency to split the molecule apart – this is known as **photodissociation**. Oxygen atoms are formed, which are radicals:

$$O_2 \xrightarrow{h\nu} 2O$$

Ozone (O_3) is formed when an oxygen atom (a radical) reacts rapidly with a dioxygen molecule:

$$O + O_2 \rightarrow O_3$$

Ozone is highly reactive. It is destroyed by reacting with radicals present in the stratosphere. If X is a radical, the two reactions involved can be written as:

$$X + O_3 \rightarrow XO + O_2 \quad \text{(equation a)}$$
$$XO + O \rightarrow O_2 + X \quad \text{(equation b)}$$
$$O + O_3 \rightarrow 2O_2 \quad \text{(equations a and b)}$$

The X produced in equation b can then continue the reactions by becoming a reactant in a repeat of equation a. Adding equations a and b together and repeating many times shows how ozone is being lost from the system.

The radical X is involved in the reaction but is not used up and so is acting as a catalyst – this is an example of a **catalytic cycle**. As a result of this, one single chlorine atom in this cycle can remove up to 1 million ozone molecules. These chlorine atoms are produced by the breakdown of chlorofluorocarbons (CFCs).

The radical X could be:
- $^{\bullet}OH$ (the hydroxyl radical) formed from water
- $NO^{\bullet}$ (nitrogen monoxide) produced in internal combustion engines
- $Cl^{\bullet}$ (chlorine radical) produced from the breakdown of CFCs which were used as cleaning solvents, refrigerants or aerosol propellants.

Why is the depletion of ozone a problem? (Storylines A2)

Ozone absorbs radiation in the region 10.1×10^{14} Hz to 14.0×10^{14} Hz. This is the ultraviolet region of the electromagnetic spectrum and is most damaging to the skin. Because much of this UV radiation is absorbed by ozone in the stratosphere, damage to the skin, such as skin cancer, is reduced. In regions of the world where the ozone layer is thinning, however, the incidences of skin cancer and other UV-related illnesses are increasing.

In the troposphere, 'ground level' ozone is an irritant toxic gas, weakening the immune system. Some ozone is formed by the action of sunlight on primary pollutants in photochemical smogs, causing breathing problems in humans.

QUICK CHECK QUESTIONS

1 Which of the following species are radicals:
   (a) OH
   (b) Br
   (c) Ne
   (d) NO_2?
2 What is the difference in the first products of heterolytic fission and homolytic fission?
3 What are the three stages of a radical chain reaction?
4 What is a chain reaction?
5 Many radical reactions take place in sunlight. What role does the sunlight play?

6 Bromine (Br_2) reacts with ethane (C_2H_6) in the presence of sunlight in a radical chain reaction. Write equations to show:
   (a) the formation of bromine radicals
   (b) the formation of C_2H_5Br
   (c) a termination reaction to form $C_2H_4Br_2$
7 What is the difference between a radical and a nucleophile? (see page 49)
8 (a) Write balanced equations to show how a chlorine radical could be involved in the removal of ozone from the atmosphere.
   (b) Why is this an example of a catalytic cycle?

Chemical equilibrium

Chemical Ideas 7.1

Dynamic equilibrium

Once a dynamic equilibrium has been established, the concentrations of reactants and products remain unchanged. However, the forward reaction and reverse reaction *do not stop* – they continue *at the same rate*.

A chemical reaction has a forward reaction, but it also has a backward reaction. If the backward reaction is significant then the reaction is **reversible**. When the rate of the forward reaction is the same as the rate of the backward reaction a system is said to be in **dynamic equilibrium**. This is represented by the symbol $\rightleftharpoons$.

For example, when carbon dioxide dissolves in water, the following reversible changes occur:

$$CO_2(g) \rightleftharpoons CO_2(aq)$$
$$H_2O(l) \rightleftharpoons H_2O(g)$$

These are physical changes.

The following reversible chemical change also occurs:

$$CO_2(aq) + H_2O(l) \rightleftharpoons HCO_3^-(aq) + H^+(aq)$$

Hydrogencarbonate ions (HCO_3^-) and hydrogen ions (H^+) form.

Steady state

Strictly speaking, a chemical equilibrium can only be established in a **closed system**. In an open system, a series of reactions can only come to a 'steady state'. An example of a steady state is the production and destruction of ozone in the stratosphere:

ozone production:	$O + O_2 \rightarrow O_3$
ozone destruction:	$O_3 \xrightarrow{h\nu} O_2 + O$
ozone destruction:	$O + O_3 \rightarrow O_2 + O_2$

None of these reactions comes to equilibrium, but left to themselves they will reach a point when ozone is being produced as fast as it is being used up, so its concentration stays the same – the series has reached a **steady state**.

Position of equilibrium

The convention is that the forward reaction for a given reaction proceeds from left to right. The backward (or reverse) reaction proceeds right to left. Reactants are always on the left of the equilibrium arrow and products are always on the right of the equilibrium arrow in a given reaction.

For any given reversible reaction there are many combinations of equilibrium mixtures possible. These combinations depend on the original concentrations of the substances and the conditions. We use the term **position of equilibrium** to describe one set of equilibrium concentrations for a reaction.

- If most of the reactants become products before the reverse reaction increases sufficiently to establish equilibrium, we say that the position of equilibrium lies to the right.

- If little of the reactants have changed to products when the reverse reaction becomes equal to the rate of the forward reaction, we say that the position of equilibrium lies to the left.

Le Chatelier's principle

The position of the equilibrium can be altered by changing the concentration of solutions, the pressure of gases or the temperature.

Le Chatelier's principle states that if a system is at equilibrium, and a change is made in any of the conditions, then the system responds to counteract the change as much as possible.

> A catalyst does not change the position of equilibrium, just the rate at which the equilibrium is established.

Concentration

Concentration change	Equilibrium shift
increasing reactant(s)	to the right (decreases reactants)
increasing product(s)	to the left (decreases products)
decreasing reactant(s)	to the left (increases reactants)
decreasing product(s)	to the right (increases products)

In the production of calcium oxide from calcium carbonate:

$$CaCO_3(s) \rightleftharpoons CaO(s) + CO_2(g)$$

carbon dioxide is removed from the kiln in order to encourage the position of equilibrium to move to the right, and increase the yield of calcium oxide.

Pressure

Pressure change	Equilibrium shift	Example
increasing	to the side with *fewer* gas molecules – in this case, to the right	$CO(g) + 2H_2(g) \rightleftharpoons CH_3OH(g)$ 3 molecules 1 molecule
decreasing	to the side with *more* gas molecules – in this case, to the right	$CH_4(g) + H_2O(g) \rightleftharpoons CO(g) + 3H_2(g)$ 2 molecules 4 molecules

Temperature

Temperature change	Equilibrium shift	Example
increase	position of equilibrium shifts in the direction of the *endothermic* reaction (in this example the equilibrium mixture becomes darker brown)	$2NO_2(g) \underset{\text{endothermic}}{\overset{\text{exothermic}}{\rightleftharpoons}} N_2O_4(g)$ brown gas colourless gas
decrease	position of equilibrium shifts in the direction of the *exothermic* reaction (in this example the equilibrium mixture becomes lighter brown)	

QUICK CHECK QUESTIONS

1 Explain what is meant by the term 'dynamic equilibrium'.

2 $C_2H_4(g) + H_2O(g) \rightleftharpoons C_2H_5OH(g)$; $\Delta H = -46\,kJ\,mol^{-1}$, $p = 70\,atm$, $T = 300\,°C$
   Predict, using le Chatelier's principle, the changes in the following that would move the position of equilibrium to the right:
   (a) concentration
   (b) temperature
   (c) pressure.

3 $Fe^{3+}(aq)$ + $SCN^-(aq)$ $\rightleftharpoons$ $[FeSCN]^{2+}(aq)$
   pale yellow   colourless   blood red
   For this reaction, what would you see if you added:
   (a) more $Fe^{3+}(aq)$
   (b) more $[FeSCN]^{2+}(aq)$?

4 Study the equilibrium reactions below. Predict the effect on the position of equilibrium of increasing the temperature.
   (a) $N_2(g) + 3H_2(g) \rightleftharpoons 2NH_3(g)$; $\Delta H = -92\,kJ\,mol^{-1}$
   (b) $N_2(g) + O_2(g) \rightleftharpoons 2NO(g)$; $\Delta H = +90\,kJ\,mol^{-1}$

5 For the equilibrium reactions in question **4**, predict the effect of decreasing the pressure on the position of equilibrium.

6 Ammonia reacts with water as follows:
   $NH_3 + H_2O \rightleftharpoons NH_4^+ + OH^-$
   (a) Suggest the likely pH of a solution of ammonia in water.
   (b) Use le Chatelier's principle to predict and explain how the position of equilibrium changes when an aqueous solution of an acid is added to the equilibrium mixture.

Rates of reaction

Chemical Ideas 10.1

> Reaction kinetics is the study of the rate of a reaction.

Rates of reaction can be affected by a number of factors:

- concentration
- pressure
- a catalyst
- intensity of radiation.
- surface area
- particle size
- temperature

Collision theory

Reactions occur when particles of reactants collide with a certain *minimum* kinetic energy:

> A particle may be an atom, ion or molecule.

- at higher concentrations and higher pressures, the particles are in closer proximity to each other encouraging more frequent collisions
- at higher temperatures, a much higher proportion of colliding particles have sufficient energy to react and more particles are able to overcome the activation enthalpy barrier
- with smaller particles of reactant there is a larger surface area on which the reactions can take place, so the greater the chance of successful collisions
- heterogeneous catalysts provide a surface where reacting particles may break and make bonds.

All of the above serve to increase the rate of a chemical reaction.

Activation enthalpy

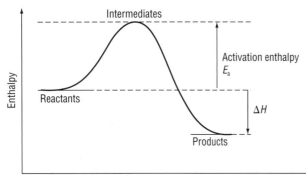

The activation enthalpy, (E_a), is the minimum kinetic energy required by a pair of colliding atoms or molecules before a reaction will occur.

An **enthalpy profile** shows how the enthalpy changes as a reaction proceeds. In this case, the reaction is exothermic.

QUICK CHECK QUESTIONS

1. Define the term 'activation enthalpy'.
2. Suggest three things you could do to increase the rate of the following reaction:

    $$Mg(s) + H_2SO_4(aq) \rightarrow MgSO_4(aq) + H_2(g)$$

3. Using the collision theory, explain why the rate of ozone depletion has increased as more CFCs have been released into the atmosphere.
4. Use the collision theory to explain why the rate of a chemical reaction is fastest at the beginning.

5. (a) Which enthalpy profile has the highest activation enthalpy, I or II?

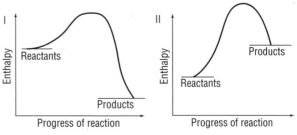

(b) Which reaction is endothermic?

The effect of temperature on rate

Chemical Ideas 10.2

Rates of reaction do not just depend on how frequently particles collide, but also on how much energy they have when a collision takes place.

Collision theory states that reactions occur when molecules collide with a certain *minimum* kinetic energy. This minimum kinetic energy is called the **activation enthalpy**. The energy needed to overcome the energy barrier is called the activation energy barrier.

As the temperature increases, the rate of a chemical reaction also increases. This is because of the distribution of energies among the reacting particles – this distribution is called the **Maxwell–Boltzmann distribution**.

To convert °C to K, add 273.
To convert K to °C, subtract 273.
e.g. 300 K = 27 °C 210 K = −63 °C

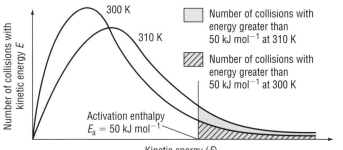

There needs to be enough molecules with sufficient energy for a reaction to take place. The molecules need to have a combined kinetic energy higher than the activation enthalpy.

Reactions go faster at higher temperatures because a larger proportion of the colliding particles have the minimum activation enthalpy needed to react.

Look back at the diagram above:

- the peak of the number of collisions at 300 K is at a lower kinetic energy value than the peak at 310 K (i.e. the most probable kinetic energy for a particle is lower at lower temperatures)
- at the kinetic energy value of 50 kJ mol^{-1} the number of collisions at 310 K is almost twice as many as at 300 K
- for reactions with an activation enthalpy around 50 kJ mol^{-1}, when the temperature rises by 10 °C (10 K) the rate of reaction approximately doubles.

QUICK CHECK QUESTIONS

1. What does E_a represent?
2. Sketch a graph showing a typical distribution of energy among the molecules of a reaction mixture.
   Shade the area of the graph representing the number of collisions with sufficient energy to lead to a reaction.
3. (a) On the graph below sketch a line showing the energy distribution for collisions occurring at $(T_1 + 10)$ K. Label this curve T_2.

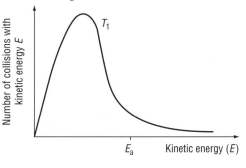

(b) Using your answer to part (a), shade the area that represents the number of collisions with sufficient energy to lead to a reaction at T_2.
4. Suggest a reason for the following observations:
   (a) N_2 and O_2 do not react at room temperature, but in a car engine the temperature is high enough for them to form NO.
   (b) The reaction between NO and O_2 to make NO_2 occurs easily at room temperature.

How do catalysts work?

Chemical Ideas 10.6

In order for any chemical reaction to proceed, bonds first need to be broken before the new bonds can be made. Bond breaking is an endothermic process – energy is taken in to break the bonds.

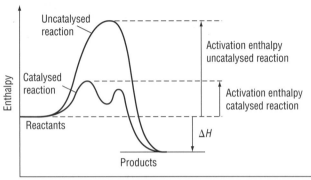

A pair of reacting molecules must collide, with a combined energy greater than the activation enthalpy for the reaction, in order to make a successful collision. Catalysts are used in order to overcome the activation energy barrier more easily. With a catalyst, successful collisions can take place at a lower energy – this is called lowering the activation energy barrier.

The reaction profiles for a catalysed reaction and an uncatalysed reaction are shown here.

Catalysts work by providing an alternative reaction pathway for the breaking and making of bonds. This alternative path has a lower activation enthalpy than the uncatalysed pathway.

Catalysts do not affect the position of equilibrium in a reversible reaction.

In heterogeneous catalysis, the reactants and catalysts are in different physical states.

Heterogeneous catalysts provide a surface on which a reaction may take place, thus lowering the energy needed for a successful collision – this lowers the activation energy barrier.

In homogeneous catalysis, the reactants and catalysts are in the same physical state.

Homogeneous catalysts work by forming an intermediate compound with the reactants. The diagram above has two humps on the lower (catalysed) pathway, one for each step.

- In the first step the activation energy barrier is overcome and an *intermediate* is formed.
- In the second step this intermediate breaks down to give a product and reform the catalyst.

QUICK CHECK QUESTIONS

1 (a) Explain why the activation enthalpy is lower for the decomposition of hydrogen peroxide solution in the presence of manganese(IV) oxide.
   (b) A solution of the enzyme catalase can also be used to decompose hydrogen peroxide solution. What *type* of catalyst is
      (i) manganese(IV) oxide
      (ii) catalase?

2 Study this enthalpy profile.

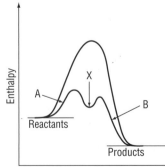

   (a) Explain, in terms of energy changes and bonds, what is happening at the points labelled A and B.
   (b) Why is there a trough (labelled X) in the pathway for the catalysed reaction?
   (c) Is the value of the enthalpy change, ΔH, affected by the use of a catalyst?

3 Ozone is decomposed in the stratosphere by chlorine atoms. The reaction takes place in two steps:

   $$Cl + O_3 \rightarrow ClO + O_2$$
   $$ClO + O \rightarrow Cl + O_2$$

   (a) ClO is formed in step 1 and removed in step 2. What do we call ClO in this respect?
   (b) Why are steps 1 and 2 together know as a 'catalytic cycle'?
   (c) Bromine atoms are about 100 times more effective than chlorine atoms at destroying ozone, despite being present in the stratosphere in much lower concentrations. What does this suggest about the reaction pathway for the reaction involving bromine?

Polymer Revolution (PR)

Many chance discoveries led to the development of the polymers we take for granted in our everyday lives. This module looks at addition polymerisation and the historical development of addition polymers. The concepts covered are given in the diagram below. 'CI' refers to sections in your *Chemical Ideas* textbook and 'Storylines' refers to *Chemical Storylines*.

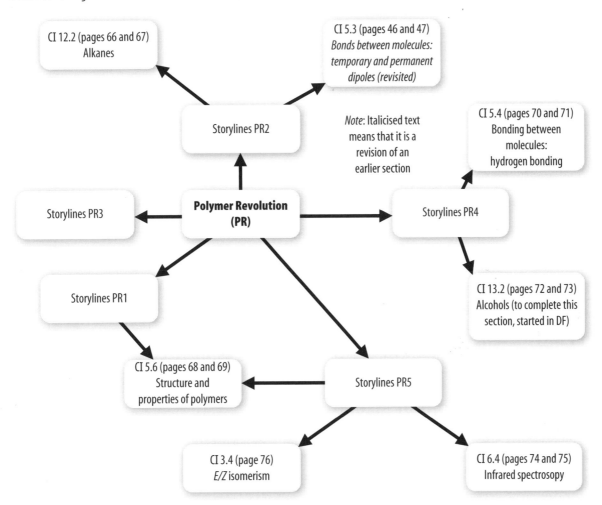

CI 12.2 (pages 66 and 67)
Alkanes

CI 5.3 (pages 46 and 47)
Bonds between molecules: temporary and permanent dipoles (revisited)

Storylines PR2

Note: Italicised text means that it is a revision of an earlier section

CI 5.4 (pages 70 and 71)
Bonding between molecules: hydrogen bonding

Storylines PR3

Polymer Revolution (PR)

Storylines PR4

CI 13.2 (pages 72 and 73)
Alcohols (to complete this section, started in DF)

Storylines PR1

CI 5.6 (pages 68 and 69)
Structure and properties of polymers

Storylines PR5

CI 3.4 (page 76)
E/Z isomerism

CI 6.4 (pages 74 and 75)
Infrared spectrosopy

From *Chemical Storylines* and the activities you will need to be aware that the uses of any polymer depend on its properties. Here are some examples of the properties and uses of some addition polymers.

Polymer	Properties	Examples of uses
LDPE	low density; flexible	carrier bags
HDPE	higher density; less flexible	buckets, food storage containers, car petrol tanks
PEX (a form of HDPE with some cross-linking)	can withstand higher temperatures than HDPE; good chemical resistance	water and gas pipes (plumbing)
PTFE	hydrophobic; slippery non-stick surface; resistant to chemical attack	'Gore-tex' clothing, 'Teflon' non-stick pans
ETFE (ethene/PTFE co-polymer)	highly transparent; low density; shatterproof; stain resistant; resistant to UV radiation	structural roofing material, such as used on the 'Eden Project' biomes
Neoprene	elastomer; non-porous; very tough; resistant to heat, light and chemical attack	wetsuits, cases for mp3 players and mobile phones

Alkenes

Chemical Ideas 12.2

> Hydrocarbons are molecules made of carbon and hydrogen *only*.

> In alk**anes** all the C–C bonds are single; alk**enes** have one or more C=C double bonds.

Alkenes are the basic hydrocarbon units of many polymers:

- –A–A–A–A–A–A– polymers are made from one type of alkene monomer
- –A–B–A–B–A–B– polymers can be made from more than one type of alkene monomer.

The general formula of alkenes is C_nH_{2n}.

Alkenes contain one or more carbon–carbon double bonds. Alkenes are said to be **unsaturated** hydrocarbons because of these double bonds.

Naming alkenes

The names of alkenes end in -*ene*. The number preceding the -ene indicates the position of the double bond. But-1-ene is $CH_3–CH_2–CH=CH_2$ and but-2-ene is $CH_3–CH=CH–CH_3$.

Alkenes can be cyclic compounds.

> When naming organic compounds, always put a dash (-) between a number and a letter and a comma between numbers.

An alkene with two double C=C bonds is called a **diene**. Note the addition of the letter 'a' after the core part of the name – for example, $CH_2=CH–CH=CH–CH_2–CH_3$ is hexa-1,3-diene.

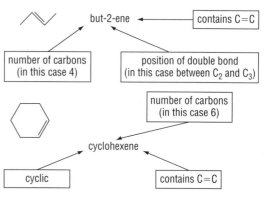

How but-2-ene and cyclohexene are named

Shape of alkenes

> Shapes of molecules and bond angles are covered on page 10.

All bond angles around the C=C double bond are 120° since there are three groups of electrons around each carbon atom (two single bonds and one double bond).

Reactions of alkenes

Alkenes undergo **electrophilic addition** reactions. In the following examples we will use ethene as a typical alkene.

There are four electrons in the double bond of ethene – these give the region between the two carbon atoms a high negative charge density.

Electrophiles are attracted to this negatively charged region in an alkene and accept a pair of electrons from the double bond at the start of the reaction.

> **Electrophiles** are either positive ions or molecules with a partial positive charge on one of the atoms. When they react they accept a pair of electrons.

A general scheme for the reactions is:

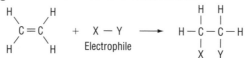

You need to know the mechanism for the addition of bromine to ethene (i.e. where X–Y is Br–Br). It can be used to demonstrate the reaction mechanism for electrophilic addition.

- When a bromine molecule approaches an alkene it becomes **polarised**.
- The electrons in the bromine molecule are repelled back along the molecule.
- The electron density is *unequally distributed*.
- The bromine atom nearest to the alkene becomes slightly positively charged.
- It now acts as an **electrophile**. A pair of electrons from the alkene moves towards the slightly charged bromine atom and a C–Br bond is formed.

- The carbon species is now positively charged; it is a **carbocation**.
- The other bromine, now negatively charged, moves in rapidly to make another bond.

The overall process is addition by an electrophile across a double bond – it is **electrophilic addition**.

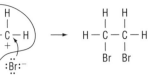

The bromine molecule becomes polarised

A carbocation

> Bromine is used as a **test for unsaturation**. The red-brown bromine becomes decolourised.

This model for the mechanism for electrophilic addition (*via* a carbocation) can be supported by experimental evidence. If chloride ions, Cl^-, are present when ethene reacts with bromine, the molecule $BrCH_2CH_2Cl$ forms as well as the expected $BrCH_2CH_2Br$. This is because both chloride and bromide ions can attack the intermediate carbocation.

Alkenes can react with a number of electrophiles. Make sure you know the reaction conditions for the following.

Electrophile	Product	Conditions
Br_2	CH_2BrCH_2Br 1,2-dibromoethane	room temperature and pressure
$Br_2(aq)$	CH_2BrCH_2OH 2-bromoethanol	room temperature and pressure
$HBr(aq)$	CH_3CH_2Br bromoethane	aqueous solution, room temperature and pressure
H_2O $(H{-}OH)$	CH_3CH_2OH ethanol	phosphoric acid/silica at 300 °C/60 atm or with conc. H_2SO_4, then H_2O at 1 atm
H_2	CH_3CH_3 ethane	Pt catalyst, room temperature and pressure or Ni catalyst at 150 °C/5 atm

> The products are also different for Br_2 as a liquid and $Br_2(aq)$ – in the latter, the water can attack the intermediate carbocation.

Addition polymerisation

Under the right conditions, alkenes can undergo addition polymerisation. The small *unsaturated* starting molecules are called **monomers** and they join together to form a long chain *saturated* **polymer**. No other product is formed.

$$CH_2{=}CH_2 + CH_2{=}CH_2 + CH_2{=}CH_2 \longrightarrow {-}CH_2{-}CH_2{-}CH_2{-}CH_2{-}CH_2{-}CH_2{-}$$

ethene (monomers) *poly(ethene)* (polymer)

This may be written as:

$$n \quad \overset{H}{\underset{H}{>}}C{=}C\overset{H}{\underset{H}{<}} \xrightarrow[\substack{\text{high temperature}\\\text{catalyst}}]{\substack{\text{gaseous phase}\\\text{high pressure}}} \left[\begin{matrix} H & H \\ | & | \\ C{-}C \\ | & | \\ H & H \end{matrix} \right]_n$$

n may be many hundreds

> The polymer is named by putting the name of the monomer in brackets and prefixing with 'poly' – e.g. choroethene monomer gives poly(chloroethene) as the polymer. Note, however, that the polymer is *not* an alkene.

QUICK CHECK QUESTIONS

1 What is the structural formula of pent-1-ene?
2 Draw a skeletal formula of the product when $H_2(g)$ reacts with pent-2-ene.
3 Name the following molecules:
   (a)   (b)   (c) $CH_2{=}CH{-}CH{=}CH_2$

4 Draw the structural formula of the product formed when HBr reacts with but-2-ene.

5 What reagent and conditions are required to convert $CH_3{-}CH{=}CH_2$ into $CH_3{-}CH(OH){-}CH_3$?
6 (a) Draw the mechanism for the reaction when Br_2 reacts with propene.
   (b) If chloride ions (Cl^-) are present when Br_2 reacts with propene, 1-bromo-2-chloropropane forms. Draw a mechanism showing how this occurs.
7 Draw out the structure of the polymer formed when the monomer $CHCl{=}CHCl$ undergoes addition polymerisation.

Structure and properties of polymers

Chemical Ideas 5.6

A **polymer** is a long molecule made up from lots of small molecules. The small molecules that add together to form a polymer are called **monomers**.

Addition polymerisation

If the monomers (represented by the letter A) contain a double bond they can add together to make a polymer:

$$A + A + A + A \longrightarrow -A-A-A-A-$$

This is called **addition polymerisation**. No other product is formed. Poly(ethene) and poly(chloroethene) are typical addition polymers.

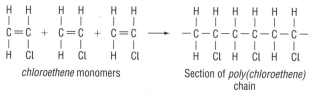

chloroethene monomers   Section of *poly(chloroethene)* chain

Examination questions often ask you to draw a small section of polymer chain, or a repeating unit.

Because the same basic unit is repeated over and over again in the polymer chain, the polymer structure can be represented more simply by drawing the **repeating unit**.

Repeating unit in *poly(chloroethene)*

Co-polymerisation

A **co-polymer** is made when two different monomers (represented here as A and B) become incorporated into the polymer chain.

$$A + B + A + B \longrightarrow -A-B-A-B-$$

In the example below, propene and ethene have co-polymerised.

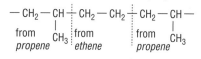

Properties of polymers

The properties of a polymer depend on six factors:

- **Chain length** – the longer the chains, the stronger the polymer. Tensile strength is a measure of how much force needs to be applied before a polymer snaps. Tensile strength increases with increasing chain length because:
  – longer chains become more entangled
  – longer chains have stronger intermolecular bonds between them and so are more difficult to pull apart.
- **Side groups on the polymer chain** – the more polar the side groups (such as Cl) the stronger the bonds between polymer chains, therefore the stronger the polymer.
- **Branching** – straight, unbranched polymer chains can pack closer together, allowing stronger bonds between the chains. This makes the polymer stronger.
- **Chain flexibility** – the more rigid the chain the stronger the polymer. Hydrocarbon chains are very flexible, whereas incorporating benzene rings makes the polymer chain stiffer.
- **Cross-linking** – more extensive cross-linking makes the polymer harder to melt.
- **Stereoregularity** – the more regular the orientation of the side groups, the closer the packing, and the stronger the polymer.

Polymers have a wide variety of properties and uses. **Elastomers** are soft and springy – they can be stretched but they return to their original shape. A well-known elastomer is rubber. Polymers that are easily molded are called **plastics** – poly(ethene) is a typical example of a plastic. **Fibres** are polymers which can be made into strong, thin threads – examples include nylon and poly(propene).

For a summary of how properties and uses of polymers relate to each other, see page 65.

Thermoplastics

These are polymers without cross-links between the chains. The intermolecular bonds between the chains are much weaker than the covalent cross-links in a thermoset.

The attractive forces in thermoplastics can be overcome by warming. The chains can slide over each other and the polymer can be deformed, i.e. change shape. On cooling, the weak bonds between the polymer chains reform and the thermoplastic holds its new shape. Examples include poly(ethene) and nylon.

Thermoplastic: no cross-linking

Weak bonds between polymer chains easily broken by heating; polymer can be moulded into new shape.

Thermosets

These polymers have extensive cross-linking between the different polymer chains. The bonds between the chains are much stronger than in thermoplastics.

The covalent bonds cannot be broken by warming. The chains cannot move relative to each other and the polymer cannot change shape. If heating continues the polymer just chars and burns. An example of this type of polymer is Bakelite.

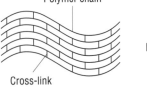
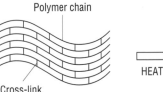

Cross-link

Thermoset: extensive cross-linking

Strong covalent bonds between polymer chains cannot be easily broken; polymer keeps shape on heating.

QUICK CHECK QUESTIONS

1 Explain what is meant by addition polymerisation.
2 Draw a section of the polymer chain formed from three monomer units of:
   (a) ethene   **(b)** but-1-ene.
3 For each of the polymers in question **2**, draw the structure of the repeating unit.
4 Which of the polymers in question **2** would be the most flexible? Explain your answer.
5 Draw the two alkenes used to make the following co-polymer:

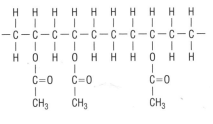

6 Draw a section of the polymer chain formed when ethene and 1,1-dichloroethene form a co-polymer.
7 What is the difference between a thermoset and a thermoplastic polymer?
8 Give three ways in which a polymer may be made stronger and less flexible.

Bonds between molecules: hydrogen bonding

Chemical Ideas 5.4

Hydrogen bonding

Hydrogen bonds are much stronger than other types of intermolecular bonds, for example instantaneous dipole–induced dipole bonds.

Hydrogen bonds have the following features:

- there is a *large dipole* between a small hydrogen atom and a highly electronegative atom (such as O, F or N)
- the *small* H atom is able to approach close to other atoms to form the hydrogen bond
- there needs to be *a lone pair of electrons* on an O, F or N atom which the hydrogen can line up with.

Hydrogen bonds occur between the molecules in liquid hydrogen fluoride. Each hydrogen atom acquires a partial positive charge because it is bonded to a highly electronegative fluorine atom. The positively charged hydrogen atom lines up with a lone pair on a fluorine atom in *another* HF molecule.

> When drawing a hydrogen bond, take care to line it up with the lone pair on an atom of O, N or F.

Hydrogen bonding in water and ice

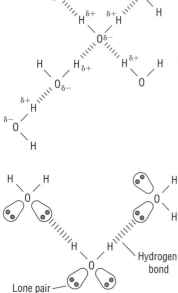

A water molecule consists of two hydrogen atoms covalently bonded to one oxygen atom. Thus the oxygen atom has two bonding pairs and two lone pairs of electrons.

Each water molecule can form twice as many hydrogen bonds as a hydrogen fluoride molecule. Water is unique in this respect and hydrogen bonding is the cause of some of water's unusual properties.

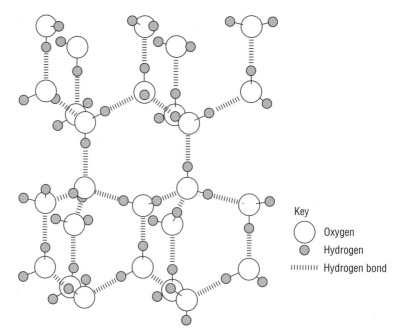

A diagram of the open structure of ice.

When water freezes, the solid ice that forms has a very open structure. This is because all the oxygen atoms are arranged tetrahedrally, with two covalent bonds to hydrogen atoms and two hydrogen bonds to neighbouring water molecules.

The effects of hydrogen bonding

Compounds with hydrogen bonding have higher boiling points than compounds with similar relative molecular masses that do not. Ethanol has a higher boiling point than propane because more energy is required to overcome the hydrogen bonds between the ethanol molecules than the instantaneous dipole–induced dipole bonds between propane molecules.

Hydrogen bonding accounts for the **solubility** of ethanol in water. Ethanol molecules form hydrogen bonds with water molecules, allowing the two liquids to mix.

Hydrogen bonding also affects the **viscosity** ('thickness') of liquids. Glycerol is an example of a viscous liquid. Each molecule has three –OH groups and this allows lots of hydrogen bonding to occur. Glycerol flows far more slowly than ethanol.

$$CH_2OH$$
$$CHOH \quad glycerol$$
$$CH_2OH$$

> When considering the relative strengths of intermolecular bonds remember the order of strength:
>
> hydrogen bonds > permanent dipole–permanent dipole bonds > instantaneous dipole–induced dipole bonds.

> Ethanol ($M_r = 46$) has a boiling point of 78 °C whereas propane ($M_r = 44$) has a boiling point of –42 °C.

Dissolving polymers

The reason why polymers such as poly(ethenol) dissolve in water comes from their structure. Poly(ethenol) has the ability to form hydrogen bonds.

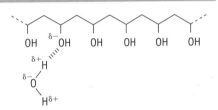
Hydrogen bonding between poly(ethenol) and water.

The –OH groups on the polymer chain can hydrogen bond with water molecules, so the polymer is soluble. The solubility can be changed by altering the proportion of –OH groups in the polymer – the polymer will become less water soluble as the proportion of –OH groups decreases. If the proportion is too large the molecule will undergo a large amount of intermolecular hydrogen bonding and it will take so much energy to pull the molecules apart that the polymer becomes virtually insoluble in water. These properties make poly(ethenol) useful for making soluble laundry bags for use in hospitals, dissolving seed coatings and dissolving stitches for use in surgery.

QUICK CHECK QUESTIONS

1 Which atoms have high enough electronegativity to form a hydrogen bond with hydrogen atoms?
2 Study the structures of the substances A–F.

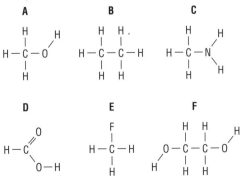

(a) Which of the substances A–F will be able to form hydrogen bonds?
(b) Which substance, A or B, has the higher boiling point?
(c) Draw a diagram showing how a hydrogen bond forms between two molecules of C.
(d) Draw a diagram to show how substance D forms hydrogen bonds with water.
(e) Which substance is likely to have the highest viscosity?
3 Why is ammonia highly soluble in water?
4 Explain why a poly(ethenol) chain that has 100% –OH groups is *insoluble* in water.

Alcohols

Chemical Ideas 13.2 (completing section started in DF)

Alcohols all contain the –OH functional group and their name ends in *-ol*. They may be classified as primary, secondary or tertiary. It is the position of the –OH group which determines the classification.

Type of alcohol	Position of –OH group	Example
primary	at end of chain $R - \overset{\displaystyle H}{\underset{\displaystyle H}{C}} - OH$	$CH_3 - \overset{\displaystyle H}{\underset{\displaystyle H}{C}} - OH$ ethanol
secondary	in middle of chain $R - \overset{\displaystyle H}{\underset{\displaystyle OH}{C}} - R'$	$CH_3CH_2 - \overset{\displaystyle H}{\underset{\displaystyle OH}{C}} - CH_3$ butan-2-ol
tertiary	attached to a carbon atom which carries no H atoms $R - \overset{\displaystyle R'}{\underset{\displaystyle OH}{C}} - R''$	$CH_3CH_2 - \overset{\displaystyle CH_3}{\underset{\displaystyle OH}{C}} - CH_3$ 2-methylbutan-2-ol

Oxidation of alcohols

You need to learn the reaction conditions for the oxidation of alcohols.

The –OH group can be oxidised using acidified potassium dichromate(VI) ($K_2Cr_2O_7$). The –OH group is oxidised to a carbonyl group. At the same time the $Cr_2O_7^{2-}$(aq) ion (which is orange) is reduced to Cr^{3+}(aq) (which is green).

During the oxidation reaction two hydrogen atoms are removed – one from the oxygen atom and one from the carbon atom.

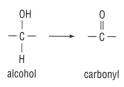

alcohol → carbonyl

The reaction conditions for the oxidation of alcohols are heating the alcohol under reflux with excess acidified potassium (or sodium) dichromate(VI) solution. For detail on reflux see the **Experimental Techniques** section (pages vi and vii)

The products of oxidation

The product depends on the type of alcohol used. With a primary alcohol an **aldehyde** is produced, which can oxidise further to give a **carboxylic acid**. The colour of the reaction mixture changes from orange to green as the dichromate(VI) ion is reduced.

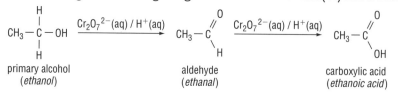

If the aldehyde is required *in situ* it can be distilled out of the reaction mixture as it is produced, in order to prevent further oxidation. If you want to make the carboxylic acid then heat under reflux with excess potassium dichromate(VI) solution.

With a secondary alcohol a **ketone** is produced and no further oxidation occurs. The colour of the reaction mixture changes from orange to green.

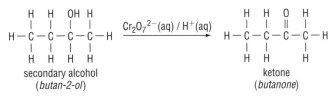

secondary alcohol
(*butan-2-ol*)

ketone
(*butanone*)

> The C=O bond is called the carbonyl bond so compounds containing a C=O group (e.g. aldehydes and ketones) are often referred to as carbonyl compounds.

Tertiary alcohols do not undergo oxidation with acidified potassium dichromate(VI) because they do not have a hydrogen atom on the carbon to which the –OH group is attached. The colour of the reaction mixture does not change, but remains orange.

You need to be able to recognise the following functional groups.

Name	Functional group
carboxylic acid	$R-C$ $\overset{O}{\underset{OH}{\diagup\diagdown}}$
ketone	$R-\overset{O}{\overset{\|}{C}}-R'$
aldehyde	$R-C$ $\overset{O}{\underset{H}{\diagup\diagdown}}$

Dehydration of alcohols

Alcohols can lose a molecule of water to produce an alkene. This is known as **dehydration** and is an example of an **elimination reaction**.

Typical reaction conditions would be using an Al_2O_3 catalyst at 300 °C and 1 atm or refluxing with concentrated sulfuric acid. An example is shown below:

> In elimination reactions, a small molecule is removed from a larger molecule leaving an unsaturated molecule. Notice that the atoms that form water are attached to *different* carbons in butan-1-ol.

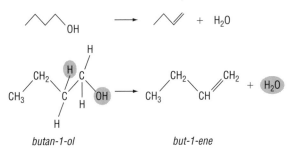

butan-1-ol

but-1-ene

QUICK CHECK QUESTIONS

1 Draw and name primary, secondary and tertiary alcohols each containing five carbon atoms.
2 Draw the structures of:
   (a) butanal
   (b) propanone
   (c) propanoic acid.
3 Draw skeletal formulae for the two oxidation products that could be obtained from the oxidation of ⌇⌇OH

4 What colour change would you observe when 2-methylpropan-2-ol was heated under reflux with acidified potassium dichromate solution? Explain your answer.
5 What reagents and conditions would you use to turn propan-1-ol into propanal?
6 Name the product from the dehydration of hexan-1-ol.

Infrared spectroscopy

Chemical Ideas 6.4

Infrared (IR) spectroscopy is a useful technique in determining the structure of organic compounds. It helps chemists to *identify different types of covalent bonds* (e.g. C=O, C=C) and hence draw conclusions about the **functional groups** in a molecule.

Why do molecules absorb IR radiation?

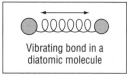

Vibrating bond in a diatomic molecule

To understand IR spectroscopy, it is helpful to think of the bond between any two atoms as being like a vibrating spring. Each bond has its own natural frequency of vibration that depends on the types of atoms forming the bond and the type of bond (single, double, triple). When a molecule is exposed to IR radiation, each bond *absorbs* energy at a particular frequency causing it to vibrate *more vigorously*. Different bonds absorb different frequencies of IR radiation.

A typical IR spectrum

The IR spectra below are for ethanol (spectrum A) and ethanoic acid (spectrum B), both in the gas phase.

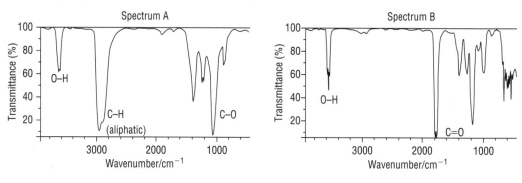

The wavenumber is the number of wavelengths that fit into 1 cm.

Every compound has a distinctive fingerprint region which can help in its identification. Absorptions in this region are caused by bonds in the molecule's 'skeleton' (e.g. C–O, C–C).

IR spectra have the following features:

- The *x*-axis shows **wavenumber**, measured in **cm^{-1}** – the scale usually starts at around 4000 wavenumbers on the left and *descends* to about 500 wavenumbers.
- The *y*-axis shows percentage **transmittance** – the baseline is at the top (100% transmittance) and the absorption signals (or bands) are *downward* troughs.
- Absorptions are described as 'strong', 'medium' and 'hydrogen bonded' (broad).
- The part of the IR spectrum below 1500 cm^{-1} is called the **fingerprint region**. Arenes often have complex absorbtion patterns in the fingerprint region.

Interpreting IR spectra

The most prominent absorption signals (bands) in an IR spectrum can be matched to a particular covalent bond using the table on page 75.

For example, in Spectrum B (above), the intense absorption in the region of 1700 cm^{-1} is characteristic of the C=O bond.

Functional groups that are involved in hydrogen bonding (e.g. –OH or –NH in alcohols, phenols, carboxylic acids and amines) usually give broad rather than sharp absorptions.

Bond	Location	Wavenumber (cm^{-1})
C–H	alkanes	2850–2950
	alkenes, arenes	3000–3100
	alkynes	about 3300
C=C	alkenes	1620–1680
⬡	arenes	several peaks in range 1450–1650
C≡C	alkynes	2100–2260
C=O	aldehydes	1720–1740
	ketones	1705–1725
	carboxylic acids	1700–1725
	esters	1735–1750
	amides	1630–1700
C–O	alcohols, ethers, esters	1050–1300
C≡N	nitriles	2200–2260
O–H	alcohols, phenols	3600–3640
	*alcohols, phenols	3200–3600
	*carboxylic acids	2500–3200

The precise position of an absorption signal depends on the environment of the bond in the molecule, so a wavenumber **region** is quoted in the table.

You don't need to learn these values. They will be given on the data sheets provided in your exam, but do familiarise yourself with these data sheets before your exam.

The asterisks refer to hydrogen-bonded groups.

Alcohol, carboxylic acid or carbonyl compound?

The presence or absence of three key IR absorptions (see table below) should allow you to distinguish between most alcohols, carboxylic acids and carbonyl compounds.

	Is there an absorption in the region 3200–3600 cm^{-1}? (caused by hydrogen-bonded –OH)	Is there a strong absorption in the region 1700–1750 cm^{-1}? (caused by C=O)	Is there an absorption in the region 1050–1300 cm^{-1}? (caused by C–O)
Alcohols, e.g. CH_3CH_2OH	yes	no	yes
Carboxylic acids, e.g. CH_3COOH	yes	yes	yes
Carbonyls, e.g. $CH_3CH_2COCH_3$	no	yes	no

QUICK CHECK QUESTIONS

1 What variables are plotted on the x-axis and y-axis in a typical infrared spectrum?

2 (a) Where is the fingerprint region in an infrared spectrum?
   (b) What part of a molecule causes absorptions in this region?

3 The IR spectra below are for two compounds, R and S, that have the same molecular formula, $C_2H_4O_2$. One is a carboxylic acid and the other is an ester.
   (a) Draw the full structural formula of each compound.
   (b) Identify the key absorptions in each spectrum and the bonds to which they correspond. Use this information to match the correct spectrum to each compound.

4 Draw the structural formulae of butanone and butan-1-ol. Identify the functional group in each molecule and give the wavenumber region in which you would expect this functional group to absorb.

5 Draw a table to show the key peaks you would expect to see in the infrared spectrum of propanoic acid, and the bond to which each absorption corresponds.

6 How does the shape of the O–H bond absorption in the infrared spectrum of *gaseous* ethanol differ from the same absorption in *liquid* ethanol? What causes the difference?

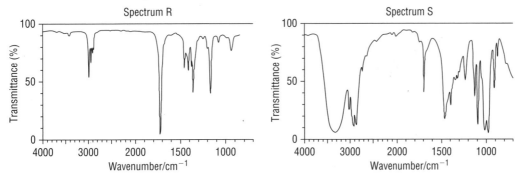

E/Z isomerism

Chemical Ideas 3.4

E/Z isomerism is one type of **stereoisomerism**. In stereoisomerism, the atoms are bonded in the same order but are arranged differently in space in each isomer.

There are two systems for naming these isomers. ***Cis–trans*** notation can be used for simple molecules whereas the ***E/Z*** notation works for both simple and complex molecules.

There are two ways of putting together the atoms of the molecule C_4H_8, but-2-ene:

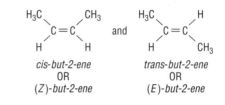

cis-but-2-ene
OR
(*Z*)-but-2-ene

and

trans-but-2-ene
OR
(*E*)-but-2-ene

> The reason why these two isomers exist is that to turn one form into the other you need to break one of the bonds in the carbon–carbon double bond. There is not enough energy at room temperature to enable this to occur. As a result, interconversion of the two isomers does not occur.

- The isomer that has the two methyl groups on the same side of the double bond is called the *cis* isomer – i.e. *cis*-but-2-ene. It is also called (*Z*)-but-2-ene.

- The isomer that has the two methyl groups on opposite sides of the double bond is called the *trans* isomer – i.e. *trans*-but-2-ene. It is also called (*E*)-but-2-ene.

QUICK CHECK QUESTIONS

1 Explain why 1,2-dibromoethane (CH_2BrCH_2Br) does not exhibit geometric isomerism.
2 Draw and label the *E* and *Z* isomers of 1,2-dibromoethene (CHBrCHBr).
3 Draw and label the *E* and *Z* isomers of pent-2-ene.
4 Explain why 2-methylpropene does not exhibit *E/Z* isomerism.

Unit F332
Practice exam-style questions

Elements from the Sea (ES)

Note: Each of these practice questions covers a single teaching module (e.g. **Elements from the Sea**). In your actual exams, each question may cover more than one teaching module – and some ideas from F331 could be re-examined, for example calculations involving amounts of substance.

1 The water in the Dead Sea is particularly rich in bromide ions, and an important chemical industry has grown up in Israel to exploit this natural resource. The main stages in the extraction of bromine from sea water are shown in the flow diagram below.

| Stage 1 | Stage 2 | Stage 3 | Stage 4 |

Dead Sea water → $Br^-(aq)$ → $Br_2(aq)$ → Impure $Br_2(l)$ → Pure $Br_2(l)$

Evaporate and acidify / Add $Cl_2(g)$ / Heat with steam

(a) (i) Describe the colour change you would expect to see in Stage 2. [2]

(ii) Explain why the process in which bromide ions, Br^-, are changed to bromine, Br_2, is referred to as *oxidation*. [1]

(b) At the end of Stage 3, the liquid bromine is contaminated with water.

(i) Give the name of the technique used to purify the bromine in Stage 4. [1]

(ii) Explain, in terms of the properties of bromine and water, how the process you have named in **(b)(i)** allows pure bromine to be separated from the mixture. [2]

(c) Special precautions must be taken when transporting bromine. Emergency breathing apparatus and protective suits must be carried by the lorry driver. Describe **two** properties of bromine that make this necessary. [2]

(d) The chlorine, used in Stage 2 to displace the bromine, is manufactured on site by the electrolysis of brine. Chlorine is produced at the anode.

(i) Write a half-equation for the production of chlorine from brine. [1]

(ii) Describe **two** other major uses of chlorine. [2]

(iii) Explain, in terms of intermolecular bonds, why bromine is a liquid at room temperature and pressure but chlorine is a gas.
QWC: In your answer, you should use appropriate technical terms, spelled correctly. [4]

> Half-equations must include electrons.
> Charges and atoms must balance.

(e) Bromine is used to manufacture the pesticide bromomethane, CH_3Br. Bromomethane is volatile so there is very little pesticide residue remaining one week after treatment.

(i) Draw the shape of the bromomethane molecule. [2]

(ii) The C–Br bond is polar. Label the diagram below to show the polarity of this bond. [1]

C—Br

> Draw in 3D, using dotted lines and 'wedges'.
> Use partial charges.

(f) The chlorine produced from the electrolysis of brine can be reacted with hydrogen gas to produce hydrogen chloride, HCl. This is dissolved in water to produce concentrated hydrochloric acid. Explain why this is a good process in terms of *atom economy*. [1]

(g) Chloroalkanes can be made in the laboratory by the reaction of alcohols with concentrated hydrochloric acid. An example is shown below.

$(CH_3)_3COH(l) + HCl(aq) \rightarrow (CH_3)_3CCl(l) + H_2O(l)$

(i) Describe how the unreacted hydrochloric acid can be removed from the reaction mixture. [3]

(ii) Name the chloroalkane produced in this preparation. [2]

[Total: 24]

> Think carefully about the practical details required – there are 3 marks to gain.

Elements from the Sea (ES)

2 Household bleach solutions contain the chlorate(I) ion, ClO^-, as the active ingredient. The chlorate(I) ion content can be determined by titration. An analyst carries out a titration as follows:

1 dm^3 = 1000 cm^3

A 5.00 cm^3 sample of bleach is reacted with an excess of acidified potassium iodide. The solution goes deep brown.

$$ClO^- + 2I^- + 2H^+ \rightarrow Cl^- + I_2 + H_2O \qquad \textbf{equation 2.1}$$

This mixture is then titrated with 0.100 mol dm^{-3} sodium thiosulfate solution, $Na_2S_2O_3(aq)$. 25.40 cm^3 of thiosulfate solution are needed to obtain the colourless end-point.

$$2S_2O_3{}^{2-} + I_2 \rightarrow S_4O_6{}^{2-} + 2I^- \qquad \textbf{equation 2.2}$$

Use volume and concentration.

(a) (i)   Calculate the number of moles of thiosulfate, $S_2O_3{}^{2-}$, used in the titration. [2]

Use the ratio from equation 2.2

    (ii)   How many moles of iodine have reacted with the thiosulfate? [1]

Use the ratio from equation 2.1

    (iii) How many moles of chlorate(I) are in the sample of bleach? [1]

    (iv) What is the concentration of chlorate(I) in the bleach, in mol dm^{-3}? [2]

(b) Calculate the concentration of the chlorate(I) ion, in g dm^{-3}, in the sample of bleach. [2]

[Total: 8]

The Atmosphere (A)

3 Carbon dioxide is one of the main greenhouse gases in the Earth's atmosphere. Other important greenhouse gases include methane, oxides of nitrogen and CFCs. Concern is increasing worldwide about the contribution these gases may be making to an enhanced greenhouse effect and global warming.

(a) Describe **two** pieces of evidence for the relationship between increased concentrations of greenhouse gases and global warming. [4]

You need a set of logical steps.

(b) Greenhouse gases absorb infrared radiation. Use this information, and your own knowledge and understanding, to explain the greenhouse effect and how it leads to warming of the Earth's atmosphere.

QWC: In your answer you should make it clear how your explanation links with the chemical theory. [6]

(c) Describe **two** possible approaches to controlling carbon dioxide emissions on a global scale. [2]

[Total: 12]

The Atmosphere (A)

4 Until recently, CFCs had been used in a variety of applications, including as propellants in aerosols. The use of CFCs was banned at the end of the last century, following scientific evidence that they were responsible for the depletion of the ozone layer in the stratosphere.

(a) (i) What does 'CFC' stand for? [1]

> Which elements must be in these compounds?

(ii) Describe **two** important properties of CFCs that make them suitable for use as aerosol propellants. [2]

> The properties must relate to the use.

(iii) Describe **two** other major uses for CFCs. [2]

(b) The ozone layer is sometimes described as the 'Earth's sunscreen'.

(i) How does the ozone layer act as the Earth's sunscreen? [2]

> You need to make two points that link how ozone behaves to the behaviour of a 'sunscreen'.

(ii) Explain what happens to CFCs in the stratosphere that leads to depletion of the ozone layer. [4]

(iii) Hydrofluorocarbons, HFCs, have largely replaced CFCs for many of their uses. Describe **one** advantage and **one** disadvantage of HFCs over CFCs. [2]

(c) Although ozone has beneficial effects in the stratosphere, it is a pollutant in the troposphere. Ozone can be formed in a complex series of reactions that leads to the formation of photochemical smogs. One of the reactions involved is

$$NO_2 \rightarrow NO + O \qquad \textbf{equation 4.1}$$

(i) The reaction in **equation 4.1** is caused by the action of sunlight on the NO_2 molecules. What type of bond breaking takes place? [1]

(ii) The oxygen atom produced in the reaction shown in **equation 4.1** is described as a radical. Explain what is meant by the term *radical*. [1]

(iii) The average bond enthalpy of the N–O bond is $+214\,kJ\,mol^{-1}$. Calculate the minimum energy (in joules) needed to break a single N–O bond. The Avogadro constant, $N_A = 6.02 \times 10^{23}\,mol^{-1}$ [2]

> Check units carefully.

(iv) In an experiment, it was found that the energy required to break the bonds in the NO_2 molecule was greater than $214\,kJ\,mol^{-1}$. Explain what this tells you about the bonds in this molecule. [1]

(v) The oxygen atom produced from the breakdown of NO_2 can react with oxygen gas in air to produce ozone. Write an equation for this reaction. [1]

(d) Nitrogen monoxide, NO, is one of a number of radicals that catalyse the breakdown of ozone in the stratosphere.

> Use the example of NO molecules in applying your general understanding of the role of catalysts.

(i) Explain how NO molecules can act as catalysts for the breakdown of ozone. [2]

(ii) Explain why the breakdown of ozone occurs faster in the hot, top layer of the stratosphere. [3]

> How does temperature affect the rate of a reaction and why?

[Total: 24]

Polymer Revolution (PR)

5 The polymer polyvinylacetate (PVA) is the main polymer used in many types of adhesives or glues.

(a) (i) The structural formula for the monomer of PVA, vinyl acetate, is shown below.

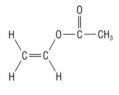

Draw the structure of the repeating unit of PVA. [2]

(ii) What **type** of polymerisation takes place when PVA is formed from vinyl acetate? [1]

> 'Thermo' means heat.

(iii) PVA is a thermoplastic polymer. Explain what is meant by the term *thermoplastic*. [2]

(b) PVA can be converted to another polymer called poly(ethenol) by replacing some, or all, of the acetate (or ethanoate) side chains with hydroxyl, –OH, groups.

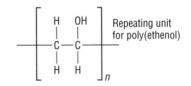

The proportion of ethanoate side chains replaced by –OH groups can be varied by changing the reaction conditions.
Some types of poly(ethenol) have the unusual property of dissolving in water.

(i) Draw a fully labelled diagram to show the intermolecular bonds in water. [4]

(ii) Suggest **one** use for a water-soluble polymer. [1]

> State the type of intermolecular bonds present as a first step.

(iii) When all the ethanoate side chains are replaced by –OH groups, the resulting polymer is referred to as 100% poly(ethenol). 100% poly(ethenol) is *insoluble* in water. Explain why.

QWC: In your answer, you should make clear how the steps in your explanation are linked. [5]

> Alcohols are classified as primary, secondary or tertiary.

(c) (i) Classify the type of alcohol group present in poly(ethenol). [1]

(ii) Describe the reagents and conditions that could be used to oxidise the alcohol group described in **(c)(i)**. [3]

(iii) Describe the colour change you would expect to see if this oxidation reaction was successful. [2]

[Total: 21]

Answers to exam-style questions

F331 Chemistry for Life

Elements of Life (EL) Question 1

Page 30

Question	Expected answers	Marks
1 (a) (i)	Two *nuclei* merge/join/combine (1); to form a bigger nucleus/different element (1)	2
(ii)	Nuclei repel each other due to positive charge/large amount of energy needed	1
(b) (i)	Atomic number and symbol correct, $_2$He (1); mass number correct, 3 (1)	2
(ii)	Isotopes	1
(iii)	Different numbers of neutrons/different mass number	1
(c) (i)	*Any two from* moderate penetrating power/stopped by a sheet of aluminium; deflected by electric fields; deflected by magnetic fields; ionising	2
(ii)	Unstable *nucleus*	1
(iii)	$\dfrac{49}{12.25} = 4$ half-lives (1) mass left $= \dfrac{100}{16} = 6.25\,\text{g}$ (1) *Correct answer alone scores 2*	2
(d)	Black lines seen (1); each element has a different/unique line spectrum (1) Spectrum from star can be compared with spectra from known elements (1) *QWC*: correct spelling and use of line spectrum (1) *Use of coloured lines/emission spectra scores 0*	4
(e) (i)	2, 5	1
(ii)	Three shared bonding pairs between N and three hydrogens (1); one non-bonding pair on N (1)	2
(iii)	with bond angle about 109°(1) Correct 3D diagram, with correct use of wedge and dashed bonds (1) Four regions of electron density (1); repel (1) each other as far apart as possible (1)	5
(f) (i)	Divide all percentages by correct A_r (1) Divide all by smallest figure (3.23) for correct ratio $1:5:1$ (1) CH_5N (1) *Correct formula alone scores 3*	3
(ii)	Highest mass peak (1) gives the M_r/(relative) molecular mass (of the compound) (1)	2
	Total	29

Developing Fuels (DF) Question 2

Page 31

Question	Expected answers	Marks
2 (a) (i)	3 C–C bonds with 9 C–H bonds attached(1); O–H bond on the end carbon (1)	2
(ii)	Alcohols	1
(b) (i)	Burns more completely/has higher octane number	1
(ii)	Made from a renewable source	1
(c) (i)	The enthalpy change when *1 mole* of a substance (1) is *completely burnt* (1) at *298 K* and *1 atm* (1)	3

(ii)	$4C(s)+5H_2(g)+1/2O_2(g) \xrightarrow{\Delta H_f^\theta} C_4H_9OH(l)$ Reactants (1); combustion products (1) $4CO_2(g)+5H_2O(l)$		2
(iii)	Hess's law $\Delta H_f^\ominus + (-2673) = 4(-394) + 5(-286)$ $\Delta H_f^\ominus = -3006 + 2673$; number 3006 (1) $= -333\,kJ\,mol^{-1}$; correct number (1) sign and units (1) *Correct answer alone scores all 3 marks*		3
(d) (i)	⌢ (1) ⌃⌄ (1)		2
(ii)	$C_3H_8(g) + 5O_2(g) \rightarrow 3CO_2(g) + 4H_2O(l)$		2
(iii)	Bonds broken $= 2E(C\text{–}C) + 8E(C\text{–}H) + 5E(O\text{=}O) = 6482$; correct number (1) Bonds formed $= 6E(C\text{=}O) + 8E(O\text{–}H) = 8542$; correct number (1) Enthalpy change = bonds broken – bonds formed = 6482 – 8542 $= -2060\,kJ\,mol^{-1}$; correct number (1) sign and units (1) *Correct answer alone scores all 4 marks*		4
(iv)	Not standard conditions/water gaseous/average bond enthalpy values used		1
(e) (i)	$24 \times 2.25 = 54$		1
(ii)	Mainly butane (1); because the M_r is close to that of butane (58) (1)		2
		Total	25

Developing Fuels (DF) Question 3

Page 32

Question	Expected answers	Marks
3 (a) (i)	The resistance to auto-ignition (1); the higher the number the more resistant (1)	2
(ii)	To prevent the engine being damaged	1
(iii)	$x = 109°$ (1); $y = 109°$ (1)	2
(b) (i)	Changes straight-chain isomers into branched-chain isomers (1); which have a higher octane numbers (1)	2
(ii)	Any correct isomer of hexane showing 6 carbon atoms (1); rest of structure correct (1); correct name (1)	3
(iii)	(Zeolite acts as a molecular sieve) which allows only straight-chain alkanes to pass through	1
3 (c)	Heterogeneous	1
(d) (i)	Heptane = straight-run; hex-1-ene = cracking; benzene = reforming; cyclohexane = reforming *All correct (2); one error (1)*	2
(d) (ii)	Hex-1-ene and benzene	1
(e)	Butane, a gas (1); in gases there are more ways of arranging the molecules or more disorder (1)	2
(f) (i)	*Any two from* carbon monoxide/carbon dioxide/sulfur oxides/particulates	2
(ii)	E.g. prevents formation of photochemical smog/inefficient/waste of fuel/irritant/causes breathing difficulties	1
(iii)	Nitrogen from air (1); reacts (with oxygen) at high temperatures in engine (1)	2
	Total	22

F332 Chemistry of Natural Resources

Elements from the Sea (ES) Question 1

Page 77

Question	Expected answers	Marks
1 (a) (i)	From colourless (1); to yellow/orange/orange red/orange brown (1)	2
(ii)	Loss of electrons/increase in oxidation state	1
(b) (i)	Distillation	1

(ii)	Bromine and water have different boiling points (1); distil at different temperatures/can be collected separately (1)	2
(c)	Bromine vapour is toxic (1); bromine is corrosive (1)	2
(d) (i)	$2Cl^- \rightarrow Cl_2 + 2e^-$	1
(ii)	*Any two from* treating (drinking) water; making bleach; making PVC	2
(iii)	instantaneous dipole–induced dipole bonds [*QWC* spelled correctly] (1) are stronger between bromine molecules (1) as the M_r is greater/larger number of electrons (1) more energy is needed to break them (so a higher b.p.) (1) *Allow reverse argument*	4
(e) (i)	Br Correct shape (1); correct use of dotted line(s)/wedge(s) to show bonds into and out of paper plane (1)	2
(ii)	$\overset{\delta+}{C} \overset{\delta-}{—} Br$	1
(f)	All the atoms (of hydrogen and chlorine produced in the electrolysis) are used to make another product	1
(g) (i)	Use a separating funnel (1); to remove the HCl layer (1) Wash with sodium hydrogencarbonate solution to remove traces of HCl (1)	3
(ii)	2-chloro-2-methylpropane: (1) for 'methylpropane'; (1) for rest correct	2
		Total	**24**

Elements from the Sea (ES) Question 2

Page 78

Question	Expected answers	Marks
2 (a) (i)	$\dfrac{25.4 \times 0.1}{1000}$ (1); $= 0.00254$ (1)	2
(ii)	Answer **(i)** $\times 0.5 = 0.00127$	1
(iii)	0.00127 *(or same answer as part **(ii)**)*	1
(iv)	Answer **(iii)** $\times \dfrac{1000}{5}$ (1); $= 0.254\,mol\,dm^{-3}$ (1)	2
2 (b)	Answer **(iv)** $\times M_r = 0.254 \times 51.5$ (1); $= 13.1\,g\,dm^{-3}$ (1)	2
		Total	**8**

The Atmosphere (A) Question 3

Page 78

Question	Expected answers	Marks
3 (a)	Historical surface-temperature records (1); show that highest temperatures have occurred in recent years (1); Modelling and predicting of past climates (1); shows that increase in recent temperatures cannot be explained by natural phenomena alone (1)	4
(b)	UV/visible radiation from the Sun (passes through atmosphere) (1); absorbed by the Earth's surface which heats up (1); Earth radiates IR (1); (Greenhouse gas molecules absorb IR) which increases vibrational energy of their bonds (1); increasing their kinetic energy and thus raising temperature (1); *QWC:* warming Earth linked to IR and warming atmosphere linked to increased vibrational energy due to absorption of IR (1)	6
(c)	*Any two from:* burn fewer fossil fuels/economise fuel use; encourage more photosynthesis/plant more trees, etc.; carbon dioxide capture (burial/reaction); use alternative (non-fossil) fuels	2
		Total	**12**

The Atmosphere (A) Question 4

Page 79

Question			Expected answers	Marks
4	(a)	(i)	Chlorofluorocarbon	1
		(ii)	*Any two from* inert; non-toxic; volatile; readily liquefied under pressure	2
		(iii)	*Any two from* refrigerant in fridges/freezers/air conditioners; blowing agent for plastic foams/polystyrene/polyurethane; dry cleaning/degreasing solvent	2
	(b)	(i)	It absorbs high energy UV radiation (1); and so protects humans/animals/plants from (cell/DNA) damage (1)	2
		(ii)	In presence of high energy/frequency UV radiation (1); the C–Cl bond breaks/undergoes photodissociation (1); to form chlorine radicals (1); that catalyse the breakdown of ozone (1)	4
		(iii)	*Advantage*: do not contain chlorine and so no chlorine radicals can be produced/break down in the troposphere (1); *Disadvantage*: act as greenhouse gases (1)	2
	(c)	(i)	Homolytic fission	1
		(ii)	(Species that) has an unpaired electron	1
		(iii)	$\dfrac{214\,000}{6.02 \times 10^{23}}$ (1) $= 3.55 \times 10^{-19}\,\text{J}$ (1)	2
		(iv)	Stronger bond (than average single N–O bond)	1
		(v)	$O + O_2 \rightarrow O_3$	1
	(d)	(i)	Reaction with NO provides an alternative route (1); which has a lower activation enthalpy (1)	2
		(ii)	Higher temperature so particles move faster/collide with greater energy (1); Higher proportion (1); of molecules collide with energy greater than the activation enthalpy (1)	3
			Total	24

Polymer Revolution (PR) Question 5

Page 80

Question			Expected answers	Marks
5	(a)	(i)	single C–C bond (1); bond at both ends (1)	2
		(ii)	Addition	1
		(iii)	On heating (1); it melts or softens (1)	2
	(b)	(i)	Water molecule correct (1); partial charges on O & H correct (1); H-bond shown by dashed line (1); H-bonded atoms in a straight line (1)	4
		(ii)	Hospital laundry bags/washing powder pouches	1
		(iii)	H-bonding (1); between poly(ethenol) molecules (1); is extensive and strong (1); too much energy required to break down (1) QWC: links between extent/strength of H-bonding and energy to overcome (1)	5
	(c)	(i)	Secondary	1
		(ii)	Acidified/dilute sulfuric acid (1); dichromate(VI) (1) [*allow formulae*]; heat under reflux (1)	3
		(iii)	Orange (1) to green/blue (1)	2
			Total	21

Answers to Quick Check questions

Unit F331: Chemistry for Life

Elements of Life (EL)

Amount of substance

Page 2

1 (a) 40.3   (b) 142.1   (c) 323.2

2 (a) CH   (b) HO   (c) $C_{12}H_{22}O_{11}$

3 (a) 0.5   (b) 0.5

4 (a) MgO   (b) $MgCO_3$

5 $BeCO_3$

6 28.8%

A simple model of the atom

Page 3

1 (a) 1p, 2n, 1e   (b) 20p, 27n, 20e   (c) 11p, 12n, 11e

2 (a) 52p, 70n, 52e (b) 95p, 146n, 95e   (c) 6p, 7n, 6e

3 (a) 6   (b) 1   (c) 2

4 80.0

Nuclear reactions

Page 5

1 (a) $^{47}_{21}Sc$ (b) $^{222}_{86}Rn$
   (c) $^{237}_{93}Np$ (d) $^{131}_{54}Xe$

2 No, the alpha particles will be stopped/absorbed by the watch

3 1 g will be left after 6.5 days, 0.5 g will be left after 13 days,
   0.25 g will be left after 19.5 days and 0.125 g will be left after
   26 days

4 Geiger counter

5 5 × 8 days = 40 days

Light and electrons

Page 7

1 (a) A black background with coloured lines on it, the lines
       gradually become closer together as the frequency increases
   (b) There are no electrons in a hydrogen nucleus

2 At least three horizontal lines; lines should become closer as the
   energy increases; two different vertical lines pointing upwards
   between levels

3 Electrons are in fixed energy levels; the electrons absorb energy
   and move to a higher energy level; the frequency of light
   absorbed is related to the difference in energy levels by $\Delta E = h\nu$

4 C is 2.4; Cu is 2.8.18.1

5 Strontium and calcium both have two electrons in their outer
   shell; these are the electrons involved in chemical reactions

6 The outer electron in potassium is further from the nucleus than
   the outer electron in sodium; the outer electron is less firmly held
   by electrostatic attraction in potassium and so is more easily lost

Chemical bonding and properties

Page 9

1 (a) covalent   (b) ionic   (c) covalent
   (d) metallic   (e) metallic

2 (a)   (b)   (c)

   (d)   (e)   (f)

3 (a) $[Li]^+$ $[\cdot\ddot{Br}\times]^-$ (b) $[Na]^+$ $[Na]^+$ $[\ddot{O}\times]^{2-}$

   (c) $[Ca]^{2+}$ $[\ddot{O}\times]^{2-}$ (d) $[Ca]^{2+}$ $[\times\ddot{Cl}\times]^-$ $[\times\ddot{Cl}\times]^-$

   (e) $[Al]^{3+}$ $[Al]^{3+}$ $[\ddot{O}\ddot{:}]^{2-}$ $[\ddot{O}\ddot{:}]^{2-}$ $[\ddot{O}\ddot{:}]^{2-}$

4 (a) ionic lattice   (b) simple molecular
   (c) covalent network   (d) simple molecular
   (e) metallic lattice   (f) ionic lattice

5 Magnesium chloride

Shapes of molecules

Page 10

1 (a) 109°   (b) 109°   (c) 109°

2 **a** = 109°, **b** = 109°, **c** = 109°, **d** = 180°, **e** = 120°, **f** = 120°

Periodicity and the Periodic Table

Page 11

1 In order of atomic number

2 The melting point increases then decreases as you go across
   Period 3

3 (a) So that they had similar properties to the group they were
       placed in
   (b) Between −220 °C and −200 °C

Group 2 and writing balanced equations

Page 13

1 $Sr(s) + 2H_2O(l) \rightarrow Sr(OH)_2(aq) + H_2(g)$

2 (a) $CaCl_2$ (b) $Sr(OH)_2$
   (c) $CaSO_4$ (d) $Ba(OH)_2$

3 Use universal indicator or another named indicator; pH > 7 or a
   colour change for the named indicator – e.g. red litmus would
   turn blue, universal indicator would turn blue or purple

4 $CaCO_3(s) \rightarrow CaO(s) + CO_2(g)$

5 (a) $Ba(OH)_2$ (b) $MgCO_3$

6 $MgO(s) + 2HCl(aq) \rightarrow MgCl_2(aq) + H_2O(l)$

7 $Mg(OH)_2(s) + H_2SO_4(aq) \rightarrow MgSO_4(aq) + 2H_2O(l)$

The mass spectrometer

Page 15

1   24.4

2

3   By an electrical field

4   Since kinetic energy = mass × velocity^2 and all ions have the same kinetic energy, heavier ions move through this region more slowly than light ions

5   **(a)** The molecule is broken down into different fragments in the mass spectrometer
    (b) CH_3COOH^+ **(c)** CH_3^+
    (d) $COOH^+$ **(e)** base peak

6   **(a)** 58   **(b)** CH_3^+ **(c)** $C_2H_5^+$

Developing Fuels (DF)

Calculations from equations

Page 19

1 **Step 1** Underline the substances whose:
     • mass you are given
     • mass you want to find.

         $\underline{4Fe} \qquad + 3O_2 \rightarrow \qquad \underline{2Fe_2O_3}$

  Step 2 Indicate moles involved   4 moles   2 moles
  Step 3 Calculate the masses   223.2 g   319.2 g

  Step 4 Convert to the mass given in the question $\frac{223.2}{223.2} \times 11.2 = 11.2\,g$

  Step 5 Convert the other mass by the same amount $\frac{319.2}{223.2} \times 11.2 = \textbf{16.0 g}$

  Step 6 Write down the answer   16.0 g of Fe_2O_3 is produced

2 **Step 1** Underline the substances whose:
     • volume you are given
     • volume you want to find

         $\underline{6CO} + \underline{13H_2} \rightarrow \qquad C_6H_{14} + 6H_2O$

  Step 2 Indicate moles involved   6 moles   13 moles
  Step 3 Calculate the volumes   144 dm^3 312 dm^3

  Step 4 Convert to the volume given in the question $\frac{144}{144} \times 6 = 6.0\,dm^3$

  Step 5 Convert the other volume by the same amount $\frac{312}{144} \times 6 = \textbf{13 dm}^3$

  Step 6 Write down the answer   13 dm^3 of H_2 is needed

3 **Step 1** Underline the substances whose:
     • mass or volume you are given
     • mass or volume you want to find

         $\underline{2NaN_3} \rightarrow \qquad\qquad 2Na + \quad \underline{3N_3}$

  Step 1 Indicate moles involved   2 moles   3 moles
  Step 2 Calculate the mass or volume   130.0 g   72 dm^3

  Step 3 Convert to the mass or volume given in the question $\frac{130.0}{130.0} \times 0.65 = 0.65\,g$

  Step 4 onvert the other mass or volume by the same amount $\frac{72}{130.0} \times 0.65 = \textbf{0.36 dm}^3$

  Step 5 Write down the answer   0.36 dm^3 of N_2 is produced

4 **Step 1** Underline
     • the mass you are given
     • the enthalpy change you are given

         $\underline{C_8H_{18}} + 12\frac{1}{2}O_2 \rightarrow 8CO_2 + 9H_2O \qquad \underline{\Delta H = -5470\,kJ\,mol^{-1}}$

  Step 2 Indicate moles involved   1 mole $\Delta H = -5470\,kJ\,mol^{-1}$
  Step 3 Calculate the mass   114.0 g

  Step 4 Convert to the mass or volume given in the question $\frac{114.0}{114.0} \times 5.7 = 5.7\,g$

  Step 5 Convert the enthalpy change by the same amount $\frac{-5470}{114.0} \times 5.7 = \textbf{-273.5 kJ mol}^{-1}$

  Step 6 Write down the answer   273.5 kJ of energy is released

5   3.58 g of hexane

6   12 150 kJ

Enthalpy and entropy

Page 20

2 (a) Copper sulfate solution has a density of $1\,g\,cm^{-3}$ and a specific heat capacity of $4.18\,J\,g^{-1}\,K^{-1}$

   (b) $-418\,kJ\,mol^{-1}$

3 Entropy increases; there are more ways of arranging gas particles *or* gases have a greater entropy than solids and liquids *or* there are more gas particles on the right-hand side

4 Entropy increases; gases have a higher entropy than liquids *or* there are more ways of arranging the particles in a gas

Hess's law

Page 21

1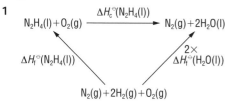

2 $\Delta H_c^{\ominus}(CO) = \Delta H_f^{\ominus}(\text{products}) - \Delta H_f^{\ominus}(\text{reactants})$
$= -393.5 - (-110.5) = -283\,kJ\,mol^{-1}$

3 $\Delta H_c^{\ominus}(N_2H_4) = \Delta H_f^{\ominus}(\text{products}) - \Delta H_f^{\ominus}(\text{reactants})$
$= 2 \times (-286) - (+51) = -623\,kJ\,mol^{-1}$

4 (a) $\Delta H_1 = \Delta H_2 - \Delta H_3$

   (b) $\Delta H_1 = -394 + 2 \times (-286) - (-890) = -76\,kJ\,mol^{-1}$

Bond enthalpies

Page 22

1 Bonds broken   Bonds made
$1 \times$ H–H $(436) = +436$ $2 \times$ O–H $(-464) = -928$
$\frac{1}{2} \times$ O=O $(498) = +249$
Total $= +685$ Total $= -928$
$\Delta H = +685 - 928 = -243\,kJ\,mol^{-1}$

2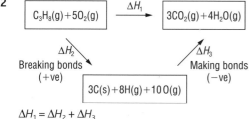

$\Delta H_1 = \Delta H_2 + \Delta H_3$

3 Bonds broken   Bonds made
$5 \times$ C–C $(347) = +1735$ $14 \times$ C=O $(-805) = -11\,270$
$1 \times$ C=C $(612) = +612$ $14 \times$ O–H $(-464) = -6496$
$14 \times$ C–H $(413) = +5782$
$10\frac{1}{2} \times$ O=O $(498) = +5229$
Total $= +13\,358$ Total $= -17\,766$
$\Delta H = +13\,358 - 17\,766 = -4408\,kJ\,mol^{-1}$

Alcohols and ethers

Page 23

1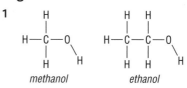
   methanol   *ethanol*

2 Pentan-2-ol

3 (a)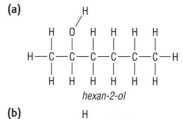
   hexan-2-ol

   (b)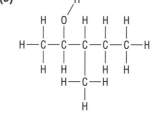
   3-methylpentan-2-ol

4 Ethoxyethane is an oxygenate so produces less carbon monoxide on burning than the corresponding alkane *or* it has a high octane number so little tendency to auto-ignite

5 OH

6 $CH_3CH_2CH_2OH + 4\frac{1}{2}O_2 \rightarrow 3CO_2 + 4H_2O$

Alkanes and other hydrocarbons

Page 25

1 Octane

2 *3-methylhexane*

   2,2,4-trimethylheptane

3 $CH_4 + 2O_2 \rightarrow CO_2 + 2H_2O$
$C_2H_6 + 3\frac{1}{2}O_2 \rightarrow 2CO_2 + 3H_2O$ or $2C_2H_6 + 7O_2 \rightarrow 4CO_2 + 6H_2O$

4 They all have the formula C_6H_{12}

5 aliphatic because it doesn't have a benzene ring

6 Cyclopropane

7 $C_{10}H_{22}$

8 $C_{10}H_{20}$

Structural isomerism

Page 26

1   (a)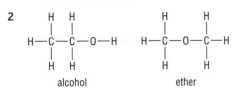

(b) They have the same molecular formula but different structural formulae

2

H—C—C—O—H   H—C—O—C—H

alcohol   ether

3   **A** and **C**

Auto-ignition and octane numbers

Page 27

1   Hexane < cyclohexane < benzene
    ──────────────────────────────→
    increasing octane number

2   Dimethylpropane; it is not an oxygenate

3   Less tendency to auto-ignite/pink/knock

Catalysts and altering octane numbers

Page 28

1   $C_8H_{18} \rightarrow C_3H_6 + C_5H_{12}$

2   Reforming (because H_2 is produced).

3   $2NO(g) + 2CO(g) \rightarrow N_2(g) + 2CO_2(g)$

Pollution from cars

Page 29

1   Incomplete combustion of a fuel/petrol

2   Contributes to formation of photochemical smog, causing respiratory problems; <u>or</u> causes acid rain which erodes buildings

3   It must be liquefied and so either low temperatures or high pressures are needed

4   Advantage: renewable/non-toxic/biodegradable/carbon neutral/lower emissions. Disadvantage: land used for growing crops for biodiesel reduces the amount of land available for food crops

5   More complete combustion <u>or</u> more reaction with air

6   Oxygenates

Unit F332: Chemistry of Natural Resources

Elements from the Sea (ES)

Ions in solids and solutions

Page 35

1. **(a)** $Na^+(aq)$ and $SO_4^{2-}(aq)$
   (b) $Ca^{2+}(aq)$ and $Cl^-(aq)$
   (c) $Fe^{3+}(aq)$ and $NO_3^-(aq)$

2. **(a)** MgF_2 **(b)** Al_2O_3 **(c)** Na_2CO_3

3. **(a)** $Pb^{2+}(aq) + SO_4^{2-}(aq) \rightarrow PbSO_4(s)$
   (b) $Ag^+(aq) + Cl^-(aq) \rightarrow AgCl(s)$

4. **(a)** $2H^+(aq) + SO_4^{2-}(aq) + 2K^+(aq) + 2OH^-(aq) \rightarrow$
   $2H_2O(l) + 2K^+(aq) + SO_4^{2-}(aq)$
   (b) Cross out all the spectator ions (those which are the same on the left and right of the arrow). This leaves the ionic equation, $H^+(aq) + OH^-(aq) \rightarrow H_2O(l)$

5. Simple cubic

6.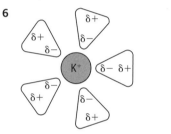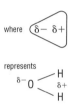

7. **(a)** barium carbonate   **(b)** lead chloride

Concentrations of solutions

Page 37

1. **(a)** $0.05\,dm^3$ **(b)** $2.5\,dm^3$

2. **(a)** 0.5 moles   **(b)** 0.4 moles

3. **(a)** $4.0\,g\,dm^{-3}$ **(b)** $196.2\,g\,dm^{-3}$

4. **(a)** $2\,mol\,dm^{-3}$ **(b)** $0.05\,mol\,dm^{-3}$

5. **(a)** $4.8\,g$ **(b)** 6.3 g

6. Conical flask, pipette (and pipette filler), burette, white tile, suitable indicator

7. **(a)** $0.02\,dm^3$
   (b) $0.02 \times 0.100 = 0.002$ moles
   (c) 0.004 moles
   (d) $0.004 \times \dfrac{1000}{25} = 0.16$ moles
   (e) $0.16\,mol\,dm^{-3}$ (2 sf)

8. $0.037\,mol\,dm^{-3}$ (3 sf)

Atoms and ions

Page 39

1. The energy needed to remove one electron from each of one mole of gaseous atoms of an element

2. **(a)** $Na(g) \rightarrow Na^+(g) + e^-$ **(b)** $C^+(g) \rightarrow C^{2+}(g) + e^-$

3. The general trend is for first ionisation enthalpy to increase across a period; this is because nuclear charge increases so there is a greater attraction between the nucleus and the outer electrons

4. **(a)** Energy must be put in to overcome the attraction of the electron for the positively charged nucleus
   (b) First ionisation enthalpy decreases down the group because the electron is further from the nucleus and is shielded by the inner electron shells

5. **(a)** Group 6   **(b)** Group 0   **(c)** Group 7

6. Magnesium has electron arrangement $1s^2\,2s^2\,2p^6\,3s^2$ and aluminum has electron arrangement $1s^2\,2s^2\,2p^6\,3s^2\,3p^1$; removing an unpaired 3p electron from aluminium takes slightly less energy than removing a 3s electron from the full 3s sub-shell in magnesium

7. **(a)**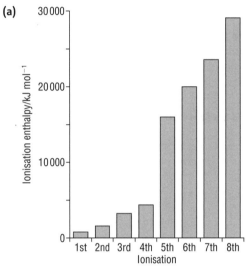
   (b) Si is in Group 4 so to remove a fifth electron would mean disrupting a full shell. This takes a lot of energy

Oxidation and reduction

Page 41

1. **(a)** K +1, Br –1   **(b)** H +1, O –2   **(c)** Co +2
   (d) P +5, O –2   **(e)** Mn +4, O –2   **(f)** Cr +6, O –2

2. **(a)** $CuCl_2$ **(b)** Cu_2O **(c)** $PbCl_4$
   (d) MnO_4^- **(e)** $NaNO_2$

3. $2Ca \rightarrow 2Ca^{2+} + 4e^-$ oxidation (electron loss)
   $O_2 + 4e^- \rightarrow 2O^{2-}$ reduction (electron gain)

4. **(a)** $Cu^{2+} + Zn \rightarrow Cu + Zn^{2+}$
   (b) $Cu^{2+} + 2e^- \rightarrow Cu$ reduction (electron gain) and
   $Zn \rightarrow Zn^{2+} + 2e^-$ oxidation (electron loss)

5. Br –1 to 0, S +6 to +4

6. **(a)** Half-equations: $3Cu \rightarrow 3Cu^{2+} + 6e^-$ and
   $2NO_3^- + 8H^+ + 6e^- \rightarrow 2NO + 4H_2O$
   Ionic equation: $3Cu + 2NO_3^- + 8H^+ \rightarrow 3Cu^{2+} + 2NO + 4H_2O$
   (b) Increase in oxidation state $= 3 \times (0$ to $+2) = +6$, decrease in oxidation number $2 \times (+5$ to $+2) = -6$

The p block: Group 7

Page 43

1. Red-brown liquid

2. **(a)** brown solution   **(b)** violet solution

3 The instantaneous dipole–induced dipole bonds between chlorine molecules are weaker because there are fewer electrons in a chlorine molecule than in a bromine molecule; chlorine molecules are therefore easier to separate from each other, taking less energy, so chlorine is more volatile than bromine

4 (a) $AgNO_3(aq) + KI(aq) \rightarrow AgI(s) + KNO_3(aq)$
   (b) A yellow precipitate would form.

5 (a) (i) $Cl_2(g) + 2Br^-(aq) \rightarrow 2Cl^-(aq) + Br_2(aq)$
         (ii) The solution would turn dark brown, due to the presence of bromine
   (b) The upper cyclohexane layer would turn red due to bromine dissolving in it.

6 Chlorine accepts electrons more easily than iodine because the atom's electrons are more strongly attracted to the nucleus in chlorine; this is because they occupy an electron shell that is closer to the nucleus in chlorine than in iodine

7 (a) Anode: $2F^- \rightarrow F_2 + 2e^-$; Cathode: $2H^+ + 2e^- \rightarrow H_2$
   (b) Oxidation occurs at the anode because fluoride ions lose electrons; reduction occurs at the cathode because hydrogen ions gain electrons

Electronic structure: sub-shells and orbitals

Page 45

1 (a) 3   (b) s, p and d

2 $1s^2 2s^2 2p^6 3s^2 3p^4$

3 (a) F   (b) Ca   (c) V

4 (a) p-block   (b) s-block   (c) d-block

5 (a) $1s^2 2s^2 2p^6 3s^2 3p^6 3d^2 4s^2$ or [Ar] $3d^2 4s^2$
   (b) $1s^2 2s^2 2p^6 3s^2 3p^1$ or [Ne] $3s^2 3p^1$
   (c) $1s^2 2s^2 2p^6 3s^2 3p^6 3d^{10} 4s^2 4p^5$ or [Ar] $3d^{10} 4s^2 4p^5$

6 (a) caesium   (b) gallium   (c) xenon

Bonds between molecules: temporary and permanent dipoles

Page 47

1 Xenon; an atom of xenon has more electrons than an atom of krypton so the instantaneous dipole–induced dipole interactions are larger so more energy is needed to pull xenon atoms apart than krypton atoms

2 In branched-chain hydrocarbons the molecules are unable to lie close together so the intermolecular forces are weaker than in close-packed straight chain hydrocarbons: the higher intermolecular forces in the straight-chain molecules means that it is more difficult to pull them apart; more energy is needed, so the boiling points are higher for the straight-chain hydrocarbons

3 At any moment in time, the electron cloud in one hydrogen molecule may be unevenly distributed, causing an instantaneous dipole in the molecule. Electrons in another nearby hydrogen molecule can be attracted (or repelled) causing an induced dipole. The positive and negative sides of these two molecules are temporarily attracted

4 Instantaneous dipole–induced dipole bonds because these occur between *all* molecules; permanent dipole–permanent dipole bonds

5 (a) Instantaneous dipole–induced dipole bonds
   (b) permanent dipole–permanent dipole bonds
   (c) permanent dipole–induced dipole bonds

6
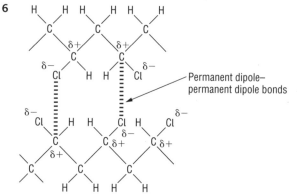

Permanent dipole–permanent dipole bonds

7 CO_2 is *not* a polar molecule; although the C=O *bond* is polar, the linear shape of CO_2 means that there is no overall dipole

Halogenoalkanes

Page 49

1 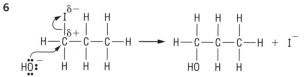 is called 4-bromo-3,3-dichloro-2-iodohexane

2 (a) CH_3Br; Br is bigger than F and has more electrons, so there are stronger instantaneous dipole–induced dipole bonds
   (b) CBr_4; it has more Br atoms in the molecule, more electrons and therefore stronger instantaneous dipole–induced dipole bonds

3 The C–I bond is weaker than the C–F bond, so it is easier to break

4 A nucleophile has a lone pair of electrons which it can use to form a dative covalent bond; examples are H_2O, NH_3, OH^-

5 Heat the halogenoalkane in a sealed tube with concentrated ammonia solution

6

7 Halogenoalkanes are immiscible in water:
   (i) separate the organic layer (halogenoalkane) and aqueous layer using a separating funnel
   (ii) shake the product with sodium hydrogencarbonate to remove any acidic impurities
   (iii) separate the product from the immiscible aqueous layer using a separating funnel
   (iv) dry the product with anhydrous sodium sulfate
   (v) purify by simple distillation

8 The R–Hal bond is broken by water

Greener industry

Page 52

1 In a batch process, reactants are added to the reaction vessel, the reaction is carried out and then the products are removed; In a continuous process, reactants are continually fed in and products taken out

2 A co-product is a product made during the desired chemical reaction, other than the target product; a by-product is a material made during an unwanted side reaction

3 **(a)** continuous   **(b)** batch

4 **(a)** The reactants that go into a production process
   (b) Hydrogen and nitrogen
   (c) Air for nitrogen; water and natural gas for hydrogen

5 The infrastructure (such as a road network) is already in place; a skilled workforce is available

6 Two of the products in the chlor-alkali industry – sodium hydroxide and chlorine – are feedstocks in bleach manufacture

7 **(a)** 100%   **(b)** 15%
   (c) the ethene is recycled

8 **(a)** Step 1 = 84.6%; step 2 = 63.1%
   (b) Recycle the HCl and reuse it in step 1
   (c) Overall 36.0%

The Atmosphere (A)

Molecules and networks

Page 55

1 Strong intramolecular covalent bonds form a giant and highly symmetrical 3D network, which is very difficult to break

2 CO_2 is made up of small molecules, which can easily be separated from each other, there is enough energy to do this at room temperature so it is a gas; SiO_2 is a solid at room temperature since it consists of a covalently bonded giant network, which needs a lot of energy to break up, so SiO_2 is a solid at room temperature

3 CO_2 dissolves in water because it has polar bonds and because hydrogen bonds can form between hydrogen atoms in water and lone pairs on oxygen atoms in CO_2; water can not break apart the giant network structure of SiO_2

4 Instantaneous dipole–induced dipole bonds *and* permanent dipole–permanent dipole bonds exist between CO_2 molecules

5 The low melting and boiling points suggest a molecular structure; the element is sulfur

What happens when radiation interacts with matter?

Page 57

1 **(a)** 0.0383%   **(b)** 0.5 ppm

2 **(a)** 2.52×10^{-20} J   **(b)** infrared   **(c)** vibrational

3 **(a)** Only certain fixed values are possible
   (b) electrons can move to a higher energy level, bonds in the molecule can break, the molecule can become ionised

4 **(a)** 8.27×10^{-19} J   **(b)** 1.25×10^{15} Hz

Radiation and radicals

Page 59

1 OH, Br and NO_2 are radicals

2 Homolytic fission produces radicals; heterolytic fission produces ions

3 Initiation, propagation and termination

4 A chain reaction continues to produce radicals which can go on to react with other molecules

5 Sunlight provides the energy for the initiation step where bonds break (homolytically) and form radicals

6 **(a)** $Br_2 \rightarrow 2Br^\bullet$
   (b) $C_2H_6 + Br^\bullet \rightarrow C_2H_5^\bullet + HBr$; then $C_2H_5^\bullet + Br_2 \rightarrow C_2H_5Br + Br^\bullet$
   (c) $C_2H_4Br^\bullet + Br^\bullet \rightarrow C_2H_4Br_2$

7 A radical has one or more unpaired electrons; a nucleophile has one or more lone pairs of electrons

8 **(a)** $Cl + O_3 \rightarrow ClO + O_2$
              $\underline{ClO + O \rightarrow Cl + O_2}$
       Overall: $O_3 + O \rightarrow 2O_2$
   (b) The chlorine radical is regenerated in the reaction and can go on to catalyse further reactions

Chemical equilibrium

Page 61

1 A reversible reaction in a closed system in which the rate of the forward reaction is equal to the rate of the reverse reaction

2 **(a)** Increase concentrations of C_2H_4 and/or H_2O; remove C_2H_5OH
   (b) lower the temperature   **(c)** increase the pressure

3 **(a)** Deepening of the red colour
   (b) Lightening of the red colour/production of yellow colour

4 **(a)** the position of equilibrium will move to the left
   (b) the position of equilibrium will move to the right

5 **(a)** the position of equilibrium will move to the left
   (b) the position of equilibrium will not shift

6 **(a)** About 8–10
   (b) H^+ ions from the acid will react with OH^- ions, removing them from the system; the position of equilibrium will move to the right to replace the OH^- ions

Rates of reaction

Page 62

1 The activation enthalpy is the minimum energy required by a pair of colliding particles before a reaction will occur

2 Increase the temperature of the reactants; use more concentrated sulfuric acid; use magnesium powder (larger surface area)

3 As the concentration of CFCs increases, there are more successful collisions between chlorine radicals and ozone; although the proportion of particles colliding with enough energy to overcome the activation enthalpy will be the same, the number of these collisions will increase.

4 At the start of a reaction the concentrations of reactants are at their highest, so there is a greater probability of successful collisions between reactant particles

5 **(a)** II   **(b)** II

The effect of temperature on rate

Page 63

1 The activation enthalpy

2

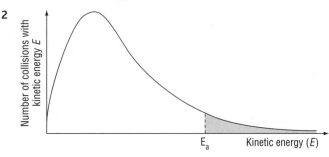

3

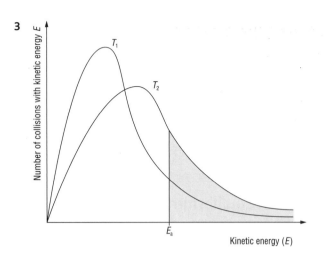

4 (a) The activation energy for the reaction between N_2 and O_2 is very high. However, inside a car engine the temperature is high enough for some molecules of N_2 and O_2 to collide with energies greater than the activation energy for the reaction, forming NO

(b) The activation energy for the reaction between NO and O_2 is low and many molecules have sufficient collision energy to exceed the activation energy at room temperature

How do catalysts work?

Page 64

1 (a) MnO_2 provides an alternative pathway for the reaction with a lower activation enthalpy; this means more particles collide with the required energy, a higher number of successful collisions occur and the rate of reaction is faster

(b) (i) heterogeneous   **(ii)** homogeneous

2 (a) A represents a point at which energy put in stretches and breaks bonds in the reactants in order to form the intermediate; B represents a point at which energy is released as new bonds form to make the final products

(b) Because an intermediate is formed in the reaction

(c) The value of ΔH does not change if a catalyst is used

3 (a) ClO is an intermediate

(b) Cl atoms are regenerated in step 2 and can go on to react with more O_3 molecules in step 1

(c) The reaction pathway involving bromine must have a much lower activation enthalpy than the reaction path involving chlorine

Polymer Revolution (PR)

Alkenes

Page 67

1 $CH_3CH_2CH_2CH=CH_2$

2 ⌇‿‿⌇

3 (a) 2,3-dimethylbut-1-ene
(b) 3-methylcyclohexene
(c) buta-1,3-diene

4

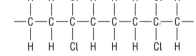

5 The reagent is H_2O; conditions: H_3PO_4 catalyst at 300 °C and 60 atm or conc. H_2SO_4 catalyst, then H_2O

6 (a)

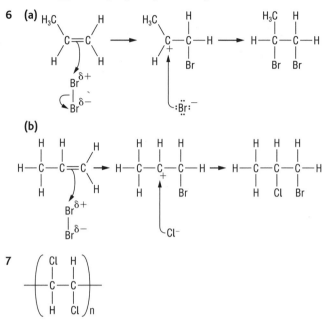

(b)

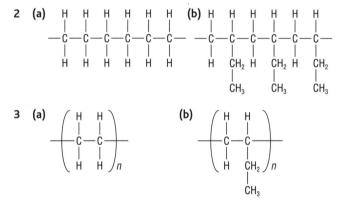

7

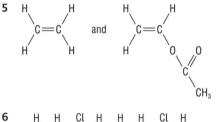

Structure and properties of polymers

Page 69

1 Small unsaturated molecules (monomers) join together to produce a long chain saturated molecule (polymer); no other product is formed

2 (a) **(b)**

3 (a) **(b)**

4 Poly(but-1-ene) would be more flexible; the chains would be further apart meaning fewer intermolecular bonds; they would slide more easily over each other

5

6

7 Thermoset plastics have cross-linking between the chains so will not melt; thermoplastics have no cross-linking so when they are heated they can be reformed; they will hold that new shape on cooling

8 Reduce branching, increase chain length, introduce polar side groups

Bonds between molecules: hydrogen bonding

Page 71

1 F, O, N

2 (a) A, C, D and F
   (b) A
   (c)

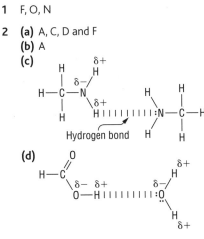

Hydrogen bond

   (d)

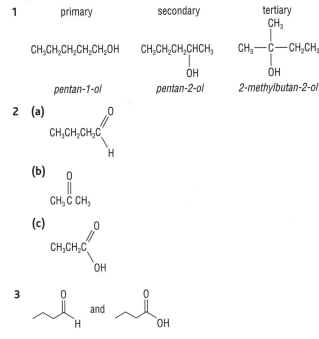

   (e) F

3 It can form hydrogen bonds with the water

4 When the polymer chain has 100% –OH groups it forms lots of intermolecular hydrogen bonds; this means that the –OH groups on the polymer are not available to form hydrogen bonds with water molecules

Alcohols

Page 73

1

primary

$CH_3CH_2CH_2CH_2CH_2OH$

pentan-1-ol

secondary

$CH_3CH_2CH_2CHCH_3$
|
OH

pentan-2-ol

tertiary
CH_3
|
$CH_3—C—CH_2CH_3$
|
OH

2-methylbutan-2-ol

2 (a)

$CH_3CH_2CH_2C$ 〈 O / H

   (b)

O
||
$CH_3 C CH_3$

   (c)

CH_3CH_2C 〈 O / OH

3

and

4 Mixture would stay orange; 2-methylpropan-2-ol is a tertiary alcohol so does not undergo oxidation under these conditions

5 Reagent: acidified potassium dichromate(VI) solution; conditions: distil off the aldehyde

6 Hex-1-ene

Infrared spectroscopy

Page 75

1 Wavenumber (cm^{-1}) is plotted on the x-axis; % transmittance on the y-axis

2 (a) The fingerprint region occurs below 1500 cm^{-1}
   (b) Absorptions in this region are caused by the 'backbone' of the molecule rather than functional groups

3 (a)

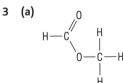

methyl methanoate   *ethanoic acid*

   (b) **Spectrum R**
   absence of a broad absorption at 2800–3200 cm^{-1} suggests no O–H group; sharp absorption at 1760 cm^{-1} suggests C=O bond; Spectrum R is methyl methanoate

   Spectrum S
   broad absorption at 2800–3200 cm^{-1} suggests O–H bond in carboxylic acid; sharp absorption at 1760 cm^{-1} suggests C=O bond; Spectrum S is ethanoic acid

4

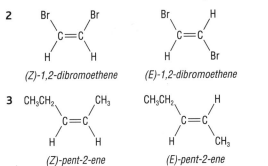

butanone

$CH_3CH_2\underset{\text{functional group}}{C}CH_3$

butan-1-ol

$CH_3CH_2CH_2CH_2\underset{\text{functional group}}{OH}$

absorbs 1705–1725 cm^{-1} absorbs 3200–3600 cm^{-1}

5

Bond	Absorption (cm^{-1})
O–H	2500–3200
C=O	1700–1725

6 In gaseous ethanol, the O–H absorption would be sharp; in liquid ethanol it would be broad because hydrogen bonding can occur.

E/Z isomerism

Page 76

1 There is free rotation about the C–C single bond

2

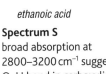

(Z)-1,2-dibromoethene   *(E)-1,2-dibromoethene*

3

(Z)-pent-2-ene   *(E)-pent-2-ene*

4 There are two hydrogen atoms at one end of the C=C double bond

Index